Designing
for Accessibility

HUMAN FACTORS AND ERGONOMICS
Gavriel Salvendy, Series Editor

Aykin, N., (Ed.): *Usability and Internationalization of Information Technology.*

Bagnara, S. (Ed.): *Theories and Practice in Interaction Design*

Buck, J.R., and Lehto, M. R., (Au): *Introduction to Human Factors and Ergonomics for Engineers*

Carayon, P. (Ed.): *Handbook of Human Factors and Ergonomics in Health Care and Patient Safety*

Hendrick, H., and Kleiner, B. (Ed.): *Macroergonomics: Theory, Methods and Applications.*

Hollnagel, E. (Ed.): *Handbook of Cognitive Task Design.*

Jacko, J.A., and Sears, A. (Eds.): *The Human-Computer Interaction Handbook: Fundamentals, Evolving Technologies and Emerging Applications.*

Karwowski, W., (Ed.): *Handbook of Standards and Guidelines in Ergonomics and Human Factors.*

Keates, S., (Au.): *Designing for Accessibility: A Business Guide to Countering Design Exclusion*

Meister, D., (Au.): *Conceptual Foundations of Human Factors Measurement.*

Meister, D., and Enderwick, T. (Eds.): *Human Factors in System Design, Development, and Testing.*

Proctor, R., and Vu, K. (Eds.): *Handbook of Human Factors in Web Design.*

Stanney, K. (Ed.): *Handbook of Virtual Environments: Design, Implementation, and Applications.*

Stephanidis, C. (Ed.): *User Interfaces for All: Concepts, Methods, and Tools.*

Wogalter, M. (Eds.): *Handbook of Warnings.*

Ye, N. (Eds.): *The Handbook of Data Mining.*

Also in this Series

HCI International 1999 Proceedings   2 Volume Set

HCI International 2001 Proceedings   3 Volume Set

HCI International 2003 Proceedings   4 Volume Set

HCI International 2005 Proceedings   11 Volume CD Rom Set
ISBN# 978-0-8058-5807-5

Designing
for Accessibility

A Business Guide
to Countering
Design Exclusion

Simeon Keates

LEA LAWRENCE ERLBAUM ASSOCIATES, PUBLISHERS

2007 Mahwah, New Jersey   London

Senior Acquisitions Editor: Anne Duffy
Editorial Assistant: Rebecca Larsen
Cover Design: Tomai Maridou
Full-Service Compositor: MidAtlantic Books & Journals, Inc.

This book was typeset in 10/12 pt Cheltenham BT, Italic, Bold, and Bold Italic. Headings were typeset in Futura Condensed BT, Bold, Italic, and Bold Italic.

Lawrence Erlbaum Associates, Inc., Publishers
10 Industrial Avenue
Mahwah, New Jersey 07430
www.erlbaum.com

Library of Congress Cataloging-in-Publication Data

Keates, Simeon.
  Designing for accessibility : a business guide to countering design
exclusion / Simeon Keates.
      p. cm.—(Human factors and ergonomics)
  Includes bibliographical references and index.
  ISBN 978-0-8058-6096-Z (case : alk. paper)—ISBN 978-0-8058-6097-9 (pbk. : alk. paper)
1. Human engineering. I. Title.
  TA166.K395 2006
  620.8'2—dc22

                                                                    2006030960

Books published by Lawrence Erlbaum Associates are printed on acid-free paper, and their bindings are chosen for strength and durability.

Printed in the United States of America
10 9 8 7 6 5 4 3 2 1

For my son, Lucian

Contents

Series Foreword

With the rapid introduction of highly sophisticated computers, (tele)communication, service, and manufacturing systems, a major shift has occurred in the way people use technology and work with it. The objective of this book series on Human Factors and Ergonomics is to provide researchers and practitioners with a platform where important issues related to these changes can be discussed, and methods and recommendations can be presented for ensuring that emerging technologies provide increased productivity, quality, satisfaction, safety, and health in the new workplace and the Information Society.

The present volume is published at a very opportune time, when the Information Society Technologies are emerging as a dominant force, both in the workplace, and in everyday life activities. In order for these new technologies to be truly effective, they must provide communication modes and interaction modalities across different languages and cultures, and should accommodate the diversity of requirements of the user population at large, including disabled and elderly people, thus making the Information Society universally accessible, to the benefit of mankind.

The present volume provides an insightful account for designing for accessibility, and the corporate benefits for designing accessibility are presented and illustrated through case studies. The book presents 14 carefully selected and very useful figures and 17 tables. The content of the book is backed up again by 88 meticulously selected references which make it very manageable to browse through them.

The book should be of special value for policy makers to ensure the practical realization of IT accessibility for all segments of the society. To corporate researchers the book indicates how and what to do for accessibility and how doing so makes good business sense. For the academic researcher it provides a framework for developing a research program in IT accessibility.

Gavriel Salvendy
Series Editor
Purdue University and Tsinghua University

Preface

This book is called "design for accessibility" because many people automatically associate "accessibility" with designing exclusively for users with disabilities and thus also with compliance with standards and legislation such as Section 508. Although compliance with legal and contractual obligations is obviously a concern, *design for accessibility* is about much more than that. It is about finding out who the users of a product or service are and designing effective solutions that meet their needs, capabilities, and aspirations.

Implementing design for accessibility within a company requires planning and forethought. Its implementation is, in effect, a business transformation process, realigning a company's product ranges to meet the needs of the users. In doing so, this offers a significant opportunity for realizing genuine innovation.

The days of products that people simply cannot use are over. Exclusion through design is no longer an option. Companies are beginning to realize that they have to move to offering products, services, and environments that do not discriminate between potential users on the grounds of capability. This is partly due to the rise in anti-discrimination legislation that is stipulating that such discrimination is no longer acceptable and the fear of lawsuits arising from this. However, that is only part of the reason companies are moving away from designing products that exclude large sections of their customer base.

Companies are looking to adopt inclusive design, universal design, etc. because they recognize economic reality. Designing for the widest possible population maximizes the number of people to whom a company can sell its products. Not only that, once one company develops a more accessible and inclusive product, all of the other products in the marketplace begin to look stale and uninviting. The ultimate reality is that companies that are selling user-unfriendly products will, in effect, be selling inferior products.

An earlier book, "Countering Design Exclusion: An Introduction to Inclusive Design," published by Springer, introduced the basic theories and ideas underlying inclusive design and the principle of countering design exclusion. This book moves beyond an introduction and focuses more on how a company can set about realizing more inclusive products. Central to the notion of a more inclusive product is, of course, that the product must offer the right combination of functionality, usability, and, equally important, accessibility. This latter attribute is a new concept for many companies and this book is intended to serve as a means to demystify what is involved in designing for accessibility.

The aim is to provide a step-by-step guide to explaining design for accessibility and help companies understand its goals and objectives and see how it can fit into their everyday design processes. This book does not attempt to provide definitive answers. There is no single correct approach to implementing design for accessibility on a companywide basis. Instead, the purpose of this book is to equip companies with the necessary frameworks and structures to help them define their own answers.

Just as if you give someone a fish, they will feed for a day, but if you teach them how to fish, they will become self-sufficient—the same principle applies here. If you teach someone an answer, they will learn that one answer, but if you provide them with a method, they can find their own answers.

Happy reading.

Simeon Keates

Acknowledgments

I would like to thank my colleagues on the BSI MS4/10 sub-committee that drafted *"BS7000 Part 6—Guide to Managing Inclusive Design,"* specifically Alan Topalian, Roger Coleman, Geoff Cook, John Gill, Lesley Morris, Anne Ferguson, Gillian Crosby, Zelda Curtis, Neil Thomas, and Jeremy Lindley.

Additionally, I would like to thank Ed Kunzinger and Andi Snow-Weaver for inviting me to join the Usable Access Council and the other members of the Council for sharing their knowledge. My thanks also go to Vicki Hanson and Stefan Hild for their support in the writing of this book and to Frances West for allowing the use of the IBM Human Ability and Accessibility model in this book and also its graphical representation on the cover.

Finally, every book has its army of proofreaders and I would like to single out Alan Topalian, Mark Laff, and Anita Keates for their eminently helpful comments and suggestions. I would also like to thank Roger Coleman, John Kemp, Todd Canterbury, and Gregg Vanderheiden for their useful and insightful reviews of this book.

Designing
for Accessibility

1

An Introduction to Designing for Accessibility

What does it take to design a *successful* product? How would you define a *successful* product? These are important questions.

Many people would answer the second question in primarily financial terms—a successful product is one that generates profit and maximizes a company's return on its investment. However, while this would be a popular answer, it is not a very useful one. Very few companies design products intentionally to make a loss on in the long run. Even if the product is intended as a loss leader, the aim is to recoup the loss sustained in future generations of the product family.

So what are the other candidate criteria for defining a successful product? Designing a product that does something new or novel and that is a first of a kind, is an alternative possible definition of a successful product. Designing a product that offers more functionality than its competitors is another option. Or perhaps, offering one that offers the same functionality, for a lower cost.

In practice, though, the most successful products are those that meet a real need or even create one that was not there beforehand. Those products find a niche in the market that they make their own.

Therefore, the secret to designing a successful product is to accurately identify a market need and then meet it. Throughout the course of this book, I will argue that there is a real market need for accessible products. Moreover, because of the design processes required to produce a product that is genuinely accessible, products that are designed to be accessible are by default well on the road to becoming successful products.

DESIGNING A SUCCESSFUL PRODUCT

In this book, the term "product" is used to refer to all consumer products, services, systems and environments that are designed for use by real people. These people

1

shall be referred to collectively as "users," unless they are likely to have paid for the product, in which case they are "customers."

Thus, designing a successful product in this extremely broad context represents a significant challenge. Many people have philosophized about how to achieve this from an engineering perspective and from a marketing point-of-view. Just look at the business section of almost any bookshop and you will see a multitude of books trying to describe the magic formula that guarantees great products and business success.

As with many things in life, there are seldom any shortcuts to be taken on the road to great products. Work needs to be performed, often from first principles, to find out what the customers and users want or need (even if they do not know it themselves). The only route to designing and developing producing great products with any real confidence of success is to adopt user-centered design practices (Mayhew, 1999). Eminent authors such as Donald Norman (1998), Ben Shneiderman (2000), and Jakob Nielsen (1993) have all published excellent works in this area. For example, in his highly influential book on usability engineering, Nielsen (1993) identified two principal attributes of successful products, namely social acceptability and practical acceptability.

Social Acceptability

Social acceptability, as defined by Nielsen and used here, is achieved when the product meets the expectations and aspirations of the end-user. We can think of these as the "user wants." Social acceptability addresses issues such as:

- Does the product look nice?
- Do I trust this product?
- Does this product stigmatize me in any way?
- Do I want this product?

This list is indicative of the type of attributes associated with social acceptability. When considering in particular the design of assistive products, it is also important to add non-stigmatizing to the list of attributes. in other words, a socially acceptable product must be one that the user is happy or content to use. Some commentators go so far as to suggest that the only successful products are those that the users want to use (Cooper, 1999). While this may be true for most so-called mainstream commercial products, it does not necessarily apply to so-called assistive or rehabilitation products, which are often bought on behalf of the user by an intermediary, such as a health authority.

Designing for social acceptability is what most product designers try to achieve.

It requires that the designers be provided with information about what constitutes a socially acceptable product for the users and then using a suitable design approach that is responsive to those requirements. Obtaining and interpreting those requirements needs specialized approaches. As would be expected, it is not possible to provide a single definition of what a user wants from a product or service. For example, some people want a product that draws attention to itself, such as a memory aid or prompt to take medication, while others would prefer something more discreet, especially for a product to be used in public. Consequently, it is often necessary to dedicate some effort to finding those wants for at least each product domain, and sometimes for each product variant.

Practical Acceptability

Nielsen's definition of practical acceptability is divided into:

- cost;
- compatibility;
- reliability; and,
- usefulness.

Of these, usefulness is subdivided further into:

- *utility*—the provision of the necessary functionality by the product to perform the desired task; and,
- *usability*—defined as including:
  - ease of learning;
  - efficiency of use;
  - ease to remembering; and,
  - low (user) error rates.

All of these attributes of practical acceptability appear eminently sensible. However, in practice, with limited time and resources allocated to the design of a new product, perhaps the ones most commonly considered the top priorities are cost and utility (also frequently referred to as functionality). After all, there is no point designing a product that is either going to cost too much to succeed in the marketplace or simply does not do what is needed of it.

It is clear to see that the ultimate cost of the product will affect its success in the marketplace; otherwise everyone would drive a Ferrari, Aston Martin, and the like. Functionality is also clearly important. A product that does not do what it is supposed to is not fit for purpose. However, functionality in certain product areas, especially computing, is in danger of becoming all-consuming. There are an increasing number of new versions of products offering essentially the same basic interface as older versions. The principal differences between the new versions and the older ones are additional functions, many of which the users may never even use. This trend to offer increasing levels of functionality simply for the sake of doing so, is known as feature creep.

Of the other attributes of practical acceptability, many are not normally considered top priorities when designing a new product, but often only middling priorities at best. Designing for reliability, for example, is a balancing act for many companies. As a consumer, I would prefer to see the products that I purchase last as long as possible without breaking down. However, it does seem that many products are designed to last the least amount of time beyond the warranty period, as possible—this is known as designing for obsolescence. Companies typically design products to last the least amount of time that they think the consumers will accept, especially in circumstances where the underlying technology is fairly mature and evolving slowly, for example: white goods such as refrigerators and washing machines.

Among those middling priorities is designing for usability. The reason for usability not being considered a top priority is that when consumers are purchasing a new product they often only consider the functions it offers and the cost of it. Usability is a property that consumers are usually only able to assess after owning the product for a period of time, after which it is usually too late to return it. After all, how many products are readily available in a shop to try before actually making a purchase? When purchasing products on-line, where one or two photographs of the product are available, the situation gets even worse for consumers who are trying to decide how easy a product may be to use.

Consequently, consumers try to reduce the amount of information that they have to process to: "How much does it cost?" and "What are the main functions it offers?" Features and price are much easier to identify and compare quickly. When buying a digital camera, for instance, customers are more likely to look for the number of megapixels and price rather than how easy it is to delete an image or put in a new memory card.

Therefore, on the whole, improved usability often will not lead directly to increased sales of a product. Thus many companies do not regard it as a top design priority. After all, what other explanation is there for the lack of really usable VCR interfaces—a technology that is over 30 years old?

However, this is often a very shortsighted view. Returning to the issue of "cost"—this has many facets. There is the cost to the company to design, build, market and support the product. There is also the cost to the consumer to buy it, the cost to maintain it and ultimately to dispose of it. A truly usable product may cost slightly more to design, but will generally cost less to support and maintain. In addition, once consumers have used a genuinely usable product, they are much more likely to remain faithful to that product's brand, rather than switch to another.

One of the best examples of this phenomenon is OXO's GoodGrips brand. Since the company was founded in 1989, its revenues have grown by over 35 percent year on year and it now boasts a range of over 500 products. This is impressive growth for a company whose products are based almost exclusively around the original concept of a usable, well-constructed handle made of durable rubber, with flexible rubber fins for improved grip. Those same distinctive fins also mark the products out on the shelf as belonging to a particular range and brand. The company attributes its success entirely to its devotion to understanding the customers' needs and practicing user-centered design (Keates and Clarkson, 2003).

Consumers are becoming increasingly sophisticated and often appreciate when a company takes that extra bit of time and trouble to ensure that a product is usable. After all, the company is effectively saying that it thinks that its customers are genuinely important people and that it is going out of its way to make sure that they are happy. Accessibility is very similar to usability in this respect.

4

DESIGNING FOR ACCESSIBILITY

The premise of this book is that the traditional concepts of utility and usability need to be extended to include a third factor to consider—that of *accessibility*. Defining accessibility in this context is not easy. Most commonly it is taken to be synonymous with designing for people with disabilities and for many circumstances this is a reasonable practical description. However, it is not a perfect or rigorous description.

Accessibility is broadly about the ability to access the functionality or utility of the product and usability is about the ability to use it. The question then is how does the ability to access differ from the ability to use?

Many of the techniques used for designing products to be accessible are basically amended or altered versions of standard usability practices. In theory, a product that is designed using user-centered design techniques and usability practices should, by default, be accessible. However, many are not accessible. Why is this?

There are several ways to think about how accessibility differs from usability, but first it helps to understand the process of interacting with a product and how the user's functional capabilities (e.g., sight, hearing, memory, hand-eye coordination, etc.) affect that. Figure 1–1 (adapted from the Center for Research and Education on Aging and Technology Enhancement's model of ageing and technology[1]) shows a basic map of the different possible interactions involved when using a product to accomplish a task within a given environment.

For an arbitrary task, say, moving from point A to point B, the user may attempt to accomplish that task without the use of a technological product. Walking to point B would be one possible option. However, if the task demands are too excessive for the user's capabilities, it may be necessary to employ a product to help accomplish the task. In this example, the product could be a car, a bus, a train or even just a walking stick. The product will have been designed to meet the demands of the task. However, the product itself will have its own set of demands that it places on the user. If those demands do not exceed the capabilities of the user, then the user is still able to accomplish the task, but if those demands exceed the user's capabilities, then the product is inaccessible and the user will not be able to perform the task.

Building on this need to balance a product's demands and the user's capabilities, consider that any design specification contains implicit assumptions about the users. It will assume their age range, level of experience with similar products or technologies, knowledge about the product and also their functional capabilities. If these properties are not defined, then the designers will typically substitute themselves as the target users—often on the basis that they are users, so why not design for the product for themselves? This is a surprisingly common attitude among designers and is what Cooper (1999) referred to as "designers design for themselves" unless directed to do otherwise.

Once the usability trials of the product start, the assumptions about who the users are often get even more entrenched into the product's operation. Any usability

---

1. http://www.psychology.gatech.edu/create/index.htm

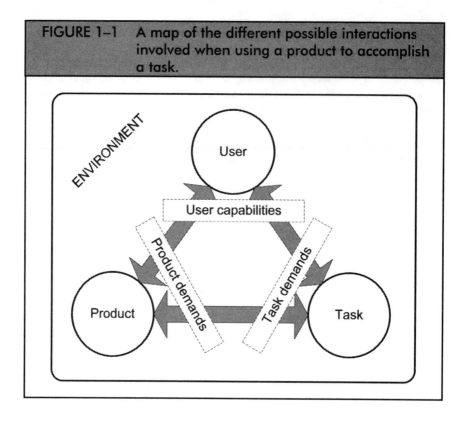

FIGURE 1-1   A map of the different possible interactions involved when using a product to accomplish a task.

trial focuses on the problems and successes identified by the particular users recruited for the trials. Any solution developed should therefore be usable and accessible by those users. Further, anyone who has similar levels of experience, knowledge, and functional capabilities to those users involved in the usability trials should be able to use the product.

However, the user recruited will often be very similar to those on whom the design specification was based. This, in itself, is not necessarily a bad thing. It means that the designers will be able to verify that the product is usable by the intended users, which is surely a good thing.

The problems arise when the assumptions about the users are examined, both for the design specification and the subsequent usability trials. Unless the assumption was framed explicitly to recognize the full range of potential user needs and capabilities it is highly likely that only a subset of needs and capabilities will be considered—typically young males in perfect health.

Thus, not all users will be similar to those involved in the user trials and those who are not, will most likely be unable to use the resultant product. There is a need to recognize those users and take steps to ensure that their needs are not overlooked. The term used to describe that need is accessibility. In other words, accessibility is the set of additional user needs not covered by the usability practices employed in the design of a particular product.

The user needs that are most likely not covered are those that arise from the particular functional limitations of specific individuals or, occasionally, across user groups. These can include slightly poorer eyesight, weaker arms, slowly declining hearing and so on. For simplicity's sake, many people immediately associate designing for accessibility with designing for people with disabilities. The reasoning for this would appear to be compelling:

- the need to design for accessibility arises from the designers not considering the full range of user capabilities when designing the product;
- thus users with limited or impaired capabilities cannot use the product and need a more accessible version to be designed;
- the user group most commonly associated with limited or impaired capabilities is the people with disabilities;
- thus designing for accessibility really means designing for the disabled.

However, this is an overly simplistic view. For example, many older adults do not possess the physical prowess of adults in their twenties. Additionally, many design teams do not consider older adults when designing products and many usability practitioners would not normally include older adults as participants in their user trials. Thus many products are designed that older adults simply cannot use because their eyesight is not quite sharp enough or their fingers not quite sufficient dexterous. Consequently, very few people would argue against the assertion that many older adults would benefit from products that have been designed accessibly. However, good luck to anyone who dares suggest to an older relative that they are "disabled."

So, perhaps designing for accessibly is about designing for the disabled and also for older adults? Well, no. Older adults are a wide and diverse set of people. Some may need help with this product, but not that one. This comment is equally true of disabled people—a deaf person typically does not need help using a calculator, for example, but may with a telephone conversation.

Furthermore, the environment plays a role in who may need additional help to use a product. The most commonly cited examples are someone on an airplane that is going through a patch of turbulence or on a train going over a junction. In such environments, even the most able-bodied person may find it difficult to move a cursor accurately onto a small icon on a computer screen or to press a tiny button on a personal CD player. So now our definition of designing for accessibility has become designing for some people with disabilities, some older adults and some people in environments that differ significantly from those in the usability laboratory.

This discussion can go on ad infinitum—women who are pregnant, someone with a temporary sports injury, such as a broken foot, someone trying to access a computer application through a mobile telephone and its smaller screen and its reduced number of keys (this scenario will become increasingly common in the next few years).

So designing for accessibility can be thought of as designing for all possible users, situations and circumstances of use. For many companies, this definition effectively equates to pushing the envelope of user capabilities and environments of use beyond those already considered in their standard design and usability practices. As such, designing for accessibility does not require the application of new skills or techniques. Instead it requires companies, their designers and usability practitioners to modify their perceptions of what the users really want and need.

This definition is not perfect, though. It is generally accepted that it is not possible to design any product that meets the needs of all users. Even something as simple in its purpose as a knife, for example, cannot be used by someone without any hands. Thus, a more pragmatic definition is that designing for accessibility is designing for all users, situations and circumstances of use that can be reasonably accommodated. The question then becomes, "What is reasonable?" The answer to that will most likely be defined in the courts through the various anti-discrimination laws that have been enacted in the past decade or so.

Legal Requirements for Accessibility

Several countries have already introduced quite stringent anti-disability discrimination legislation or are planning to do so in the very near future. These laws typically began as a means of ensuring equal civil rights for all sections of the population. However, over time these laws have been evolving, or been added to, and now are beginning to specify accessibility targets for particular technologies, products and services. Under this newer type of legislation, the onus is on a company to justify why making a product accessible represents an unreasonable burden. To make the situation more complex for companies, the definition of "unreasonable burden" has yet to be established in the courtroom and, indeed, may vary from country to country. Equally, different countries have their own interpretations of what should be accessible.

Please note that what follows is not a complete discussion of all of the applicable legislation pertaining to accessibility. It would take a highly skilled lawyer to provide that. Instead, the aim is to highlight the different types of legislation with which a company and its product offerings may have to comply.

In the United States, there are a number of statutes of which companies need to be aware (U.S. DoJ, 2004). The first of these was the 1973 Rehabilitation Act (Rehab Act, 1973), which aimed to establish a baseline for prohibiting discrimination on the grounds of disability in programs conducted by Federal agencies or receiving Federal financial assistance, in Federal employment, and in the employment practices of Federal contractors. This Act was amended in 1992 and again in 1998.

The 1990 Americans with Disabilities Act (ADA, 1990) is a piece of civil rights legislation that prohibits discrimination against people with disabilities in employment, State and local government, telecommunications, commercial facilities, transportation, and public accommodations. This latter requirement, public accommodation, is open to interpretation with respect to web sites. Legal opinion appears divided as to whether the Web is a public space and therefore subject to the ADA, although it does seem that the argument is slowly moving towards accepting that it is.

Section 255 of the 1996 Telecommunications Act (Telecoms Act, 1996) represents a shift in objective from ensuring that U.S. citizens with disabilities are not prevented from exercising their civil rights, to one where the emphasis is on ensuring that they have equal access to a technology. Section 255 stipulates that all telecommunications equipment manufactured or sold in the U.S. is:

> ". . . designed, developed, and fabricated to be accessible to and usable by individuals with disabilities, if readily achievable."

Section 255 is fundamentally different to the ADA and Rehabilitation Act in that a company can be found to be in breach of it without a complaint having to be filed by someone who feels discriminated against.

Perhaps the most original and innovative piece of legislation is known simply as Section 508. Section 508 of the 1973 Rehabilitation Act was amended in 1998 in what became the Reauthorized Rehabilitation Act (confusingly also known as the Workforce Investment Act—WIA, 1998). This legislation prohibits the U.S. Federal Government and all of its agencies from purchasing, using, maintaining or developing any electronic and information technology (E & IT) products that are not deemed fully accessible.

At first sight, this might not seem to be of great importance for many companies. After all, this legislation only refers to U.S. government purchases. However, Section 508 is significantly more influential than that. For a start, the U.S. government is currently the world's biggest purchaser of IT products. Very few companies in the IT arena would deliberately choose to ignore such a huge market. Consequently, in the United States at least, the accessibility requirements specified in Section 508 are becoming de facto standards in the IT industry. Those requirements are being codified into checklists that IT companies simply cannot afford to ignore. Furthermore, other governments around the globe, at the national and local levels, have seen the success of Section 508 and are actively investigating their own local variants.

Internationally, there are few governments that have been as active in implementing anti-discrimination legislation as the United States has. In Canada, discrimination is outlawed through the Canadian Human Rights Act (CHRA, 1985), although this only applies to:

- federal departments, agencies and Crown corporations;
- the post office;
- chartered banks;
- airlines;
- television and radio stations;
- inter-provincial communications and telephone companies;
- buses and railways that travel between provinces; and,
- other federally-regulated industries.

Filling in the gaps are provincial or territorial human rights acts or codes, which all have blanket anti-discrimination provisions (Cleveland, 2002).

In Europe, the European Union has been contemplating specific anti-discrimination legislation, but has so far shied away from implementing anything beyond the

EU Charter of Fundamental Rights (Europa 2000) and Article 13 of the EC Treaty (Europa, 2002). However, that has not stopped member nations from implementing their own legislation in this area. The United Kingdom, for example, introduced the Disability Discrimination Act (DDA, 1995), which ensures the rights of people with disabilities with regard to:

- employment;
- education—also addressed through the 2001 Special Educational Needs and Disability Act (SENDA, 2001);
- access to goods, facilities and services—including shop premises;
- buying or renting land or property; and,
- public transport—the UK government has the power to set minimum standards so that people with disabilities can use public transport easily.

To help companies understand their obligations under the DDA and also to ensure consistent interpretation of those obligations, the UK government established the Disability Rights Commission (DRC). The DRC is responsible for compiling Codes of Practice relating to the various Parts of the DDA. The Codes of Practice (DRC, 2005) provide practical examples of how the DDA should be interpreted and will form the basis of future legal judgments.

Outside of Europe, Australia, for example, implemented its own Disability Discrimination Act back in 1992 (Aus DDA, 1992), which is overseen by the Australian Human Rights and Equal Opportunity Commission. The DDA (Australia) addresses very similar issues to those of its UK namesake, with the notable additions of access to sports and clubs or societies.

It is quite clear that there is little consistency internationally on how to legislate on anti-discrimination issues. Countries such as the United States have a wide range of laws dealing with this issue, while many others prefer to issue directives rather than enact legislation. It should, therefore, come as no surprise that international attitudes to accessibility also vary widely.

INTERNATIONAL ATTITUDES TOWARDS ACCESSIBILITY

In the United States, attitudes toward people with disabilities have been led largely by the civil rights background to anti-discrimination legislation. The ADA has been structured around the principles of similar legislation addressing racial discrimination. Thus concepts such as the unacceptability of "equal, but different" access are enshrined in law. A practical example of this is a wheelchair ramp leading to a public building. In theory, access to the building can be achieved by positioning the ramp leading to any entrance to the building. Under such circumstances, the building would be accessible. However, under the ADA, if the ramp does not lead to the same main public entrance that people who are not wheelchair-users would use, then it would fail to meet the requirements of the ADA because it would constitute "different" access.

Extending this concept of people with disabilities are, first and foremost, equal citizens who happen to have particular functional limitations, the terminology used in the United States has developed to be person-first. Thus, the user group of interest here is people (or persons) with disabilities and not disabled people, a person with a visual limitation and not a blind person, and so on.

In Europe, the situation is somewhat different. According to a survey by the BBC Ouch magazine, "disabled people" is preferred to "people with disabilities" among disabled respondents to a survey in the UK (Rose, 2004) and most UK government publications now use that phrase. There are numerous reasons for this transatlantic divide. First, many European countries have a significantly more widespread welfare state and corresponding government benefits than in the United States. Consequently, the situation is much less clear-cut as to whether a "label" is disadvantageous and derogatory or not. Being "labeled" as "disabled" in the UK, for example, entitles a person to additional governmental benefits, such as an Attendance Allowance or Disability Living Allowance.

However, in themselves, such financial incentives do not fully explain the use of the "disabled people" rather than "people with disabilities." A better explanation is offered by the BBC survey mentioned above. It claims that the phrase "people with disabilities" is based on the presumption that the disability is a characteristic or attribute of the person and that "disabled" is somehow a dirty word. The thinking behind the phrase "disabled people" is that the adjective "disabled" is a property of society and the environment and not the person. Consequently, a person is not fundamentally disabled as such, but rather is disabled by external factors. Thus, if the environment was designed to be accessible, the person would cease to be disabled. Under those circumstances, where disability is not an attribute of the person but of an environment that is "disabling," then the term "disabled" should not be viewed as a negative attribute of the person, but of society's response to that person's needs.

Finally, there is also a view that "people with disabilities" is almost patronizing to some people, just as the terms "special" and "brave" come in for particular criticism from the survey respondents. This is not to say that the phrase "people with disabilities" should not be used. In the United States a person with a disability may well be offended at being referred to as a "disabled person" and so in that situation the former phrase should most definitely be used instead of the latter. The purpose of this discussion was to highlight that there is no globally acceptable terminology and that the reason for this is as much a result of cultural differences as anything else. Those differences have not only led to differences in terminology, but also to different approaches to designing accessible products.

INTERNATIONAL APPROACHES TO DESIGNING
FOR ACCESSIBILITY

Just as legislation and governmental responses vary from country to country, so too do the approaches to ensuring that all products are as accessible as possible. This is no great surprise, as the cultural factors that led to the adoption of specific laws within particular countries also play an influential role in how those countries tackle the issue.

Universal Design

Universal design is probably the pre-eminent design philosophy in the United States and Japan that addresses the accessibility of products. It is defined as:

> ". . . the process of creating products (devices, environments, systems, and processes) which are usable by people with the widest possible range of abilities, operating within the widest possible range of situations (environments, conditions, and circumstances), as is commercially practical."
>
> (Vanderheiden and Tobias, 2000)

Further, universal design is underlined by 7 guiding principles (Follette Story, 2001):

1. **Equitable use**—the design is useful and marketable to people with diverse abilities.
2. **Flexibility in use**—the design accommodates a wide range of individual preferences and abilities.
3. **Simple and intuitive**—use of the design is easy to understand, regardless of the user's experience, knowledge, language skills, or current concentration level.
4. **Perceptible information**—the design communicates necessary information effectively to the user, regardless of ambient conditions or the user's sensory abilities.
5. **Tolerance for error**—the design minimizes hazards and the adverse consequences of accidental or unintended actions.
6. **Low physical effort**—the design can be used efficiently and comfortably and with a minimum of fatigue.
7. **Size and space for approach and use**—appropriate size and space is provided for approach, reach, manipulation, and use regardless of user's body size, posture, or mobility.

For more detailed descriptions and guidance on universal design, the Web sites of the Center for Universal Design at North Carolina State University[2] and the Trace Center at the University of Wisconsin-Madison[3] are excellent resources.

Readers should note the implicit prohibition of relying on bolt-on assistive devices or technology to make a product accessible in the definition of universal design. In a world where every product was designed according to universal design principles, assistive technology would become unnecessary—all products would be accessible as-is, off the shelf. This is patently a laudable goal; however it is beset by pragmatic limitations, principally technological.

---

2. http://www.design.ncsu.edu/cud/
3. http://trace.wisc.edu/

Despite all the wondrous advances in technology over the past hundred or so years and, in particular, since the advent of the computer, it is difficult to imagine a machine that is capable of being able to offer a user exactly the functionality that the user wants *all of the time*. There may be some that come very close to this, for instance machines that perform a very limited set of functions, but even these, on occasion, will frustrate the user. It is also arguable that those machines only succeed because the users have generally become familiar with the functionality of them and will generally seek alternatives if they believe the machine will not offer a particular function that they are looking for.

An example of this is a ticket machine in a train station. Most commuters will know exactly what ticket they want to buy and the ticket machine would appear to meet their needs precisely. However, what about the night of the office Christmas party? If a commuter decides that he or she would like to stay the night in a hotel in the city instead of commuting home that evening, is the ticket machine likely to offer the ability to buy a ticket to the city for today, but a return for tomorrow evening? This circumstance is most likely beyond the options offered by the machine and the commuter is likely to go to the ticket office to buy the necessary tickets there.

The message of this example is that if a machine has not been designed to cope with a very specific set of circumstances, then it is likely that when those circumstances arise, the machine will not be able to rise to the challenge. Thus, the only way of designing any machine or product that will be accessible to *all* users is to consider the needs of *all* users throughout the design process. For it to be accessible for *all* users in *all* possible circumstances, the complexity of the design space begins to grow exponentially.

If we add to that the strong likelihood that the needs of some users in some circumstances will be diametrically opposed to other users in the same or different circumstances, it should be apparent that universal design as stated is virtually impossible to achieve. Thus, as a utopian ideal, universal design should be applauded, but it will most likely remain out of reach as an attainable goal for the foreseeable future. In light of this, one might question why universal design has such a bold goal. The reason is most likely the cultural origin of it. Universal design originated in the United States in response to anti-discrimination measures, such as the ADA, which, in turn, are based on the utopian ideal of equitable access for all and that there should be no difference in how the access is attained.

In Europe, where anti-discrimination legislation has come about from a different set of cultural mores, universal design has been less well received, and inclusive design and design for all appear to be the preferred philosophies.

Inclusive Design

As already mentioned, universal design is not universally accepted to be the best guiding philosophy for ensuring satisfactory levels of access for all. In Europe, for example, inclusive design is arguably the preferred philosophy. Its definition is not as tightly defined (and thus constrained) as that of universal design. However, that does not mean that definitions do not exist. For example, inclusive design is defined by the UK Department of Trade and Industry (DTI) as a design goal for which:

> *". . . designers ensure that their products and services address the needs of the widest possible audience."*
>
> (DTI Foresight, 2000)

This definition implies the creation of products and services is the central focus of inclusive design. Meanwhile, the Royal Society for the Encouragement of Arts, Manufactures and Commerce (otherwise known as the RSA), defines inclusive design more broadly as being:

> *". . . about ensuring that environments, products, services and interfaces work for people of all ages and abilities."*
>
> (RSA, 2005)

The UK Design Council has a further definition:

> *"Inclusive design is not a new genre of design, nor a separate specialism, but an approach to design in general and an element of business strategy that seeks to ensure that mainstream products, services and environments are accessible to the largest number of people."*
>
> (Coleman, 2005)

While the lack of a single definitive definition may initially be regarded as a weakness, many practitioners and researchers regard it as a strength. It means that the definition does not rule out potential solutions for achieving accessibility simply on the grounds of an ideological stance.

There are common themes within the definitions that are attractive to companies looking to make their products more accessible. For instance, the definitions are all based on the idea of making so-called mainstream products (i.e., those not designed explicitly for users with disabilities) as accessible as possible for as many potential users as possible. There is no commitment here to "one product for all." Instead there is recognition that, with the best intentions in the world, it is virtually impossible to design even a simple product that absolutely everyone can use. There will most likely always be extreme users with such unique needs that only a custom-built solution will suffice. The aim of inclusive design is to minimize the number of users requiring such solutions. Many companies consider this to be a much more realistic, albeit less idealistic, view of the world than universal design and its requirements of access for *all* users.

Design for All

Design for all has unfortunately become synonymous with "one product for all." in practice, most available definitions from design for all advocates are surprisingly similar to those for inclusive design (and therefore *not* one product for all). For example, the European Union's eEurope initiative defines it as:

> "... *designing* mainstream *products and services to be accessible by as broad a range of users as possible.*"

(Europa, 2005)

The European Union goes further and sponsors the Design for All and Assistive Technology (DfA & AT) Awards (DfA:AT, 2005), which are given out at a single joint ceremony. This is an explicit recognition that the two approaches are complementary and are needed together to provide accessible solutions for all potential users.

Countering Design Exclusion

Design exclusion is defined by the new British Standards Institution publication "BS7000 Part 6—Guide to managing inclusive design" as:

> "*inability to use a product, service or facility, most commonly because the needs of people who experience motor, sensory and cognitive impairments have not been taken into account during the design process.*"

(BSI, 2005)

Knowing who and how many people cannot use the product, and why they cannot do so, immediately highlights the aspects of the product that need to be improved. For example, if a product excludes a significant proportion of the population because the users either cannot hear or see the output from the product, then designers know to re-design the features involved in providing the output to the users.

The underlying principle of countering design exclusion is that by identifying the capability demands placed upon the user by the features of the product, it is possible to identify users who cannot use the product. A person's ability to use a product relies on his or her capabilities (e.g., strength, ability to see, etc.) meeting or exceeding those required to use the product. Importantly, this ability to use the product or not is determined primarily by each user's capabilities and are irrespective of the cause of each user's functional impairment(s). The ultimate goal of countering design exclusion is to re-design the product to lessen the demands placed upon the user, so that a wider range of users can potentially be included and no one is excluded unnecessarily.

An analogy of the differences between inclusive design and countering design exclusion can be found in the world of human-computer interaction (HCI) and the methodologies for evaluating how easy it is to click on a particular graphical user interface entity, such as, for instance, an icon. Traditional HCI techniques are based around models and rules such as Fitts' Law (Fitts, 1954) which focus on describing successful tasks. In the case of Fitts' Law the description is a mathematical equation describing the relationship between the size of a target, the distance to the target from the starting point and a correcting coefficient known as the index of difficulty to the time taken to reach the target. Any errors that arise, say, by missing the target or pressing the wrong mouse button, are simply logged but not analyzed.

However, under the principles of countering design exclusion, the errors are the most interesting and useful part of the data. Knowing, for example, that certain users miss small targets because the target is smaller than the cursor is a very useful fact to know. Finding out that the reason for this is because the users were never told that it is the tip of the cursor arrow (on many graphical user interfaces) that is the active part of the cursor and not the center of the arrow, is even more useful.[4] Thus the designer could make the interface easier to use for those users by showing when the active part of the cursor is over a target, through changing the appearance or color of the target or adding a beep. Incidentally, this solution would also benefit users with poor eyesight who may not be able to see exactly where the tip of the cursor is. Approaches based purely on analyzing the successful tasks would not provide the insight necessary to develop such a straightforward solution.

This example shows that countering design exclusion is entirely consistent with, and complementary to, inclusive design and design for all. It can best be thought of as a practical approach for achieving them.

KEY POINTS

- Designing a successful product requires finding and meeting a market need.
- All successful products need to offer utility, usability, and accessibility.
- Accessibility requirements vary from country to country.

---

4. I have encountered such users frequently over the years. They are most commonly older adults who have only recently learned to use computers. Most training courses assume that the novice user already has this type of knowledge, but in practice many of them do not.

2

Making the Business Case for Accessibility

In the previous chapter, we examined the legislative imperatives for companies to adopt design for accessibility. However, basing a business case for accessibility on legislative requirements is not a good strategy for generating truly accessible products. The reason for this is that most legislation is based on the concept of products having to meet a minimum level of accessibility to avoid potential lawsuits. This in turn leads to many companies adopting the attitude that if the only reason they are considering making their products accessible is to avoid such litigation, then there is no incentive to go beyond the minimum levels of accessibility stipulated by the legislation.

Consequently, basing the business case solely on the legislative requirements does promote the argument for accessibility, but only so far and then brings it to a grinding halt again. The net effect for potential customers is that the situation has improved for them (the products are more accessible), but companies are, most likely, still not designing the "killer" product or application that they are looking for, the one that genuinely meets all of their needs. The question then arises—what other arguments are there that promote the business case for accessibility? We will examine a number of possible answers to this question in the rest of this chapter.

"THE POPULATION IS GETTING OLDER"

In virtually every country in the so-called developed world, the population is ageing and ageing rapidly. Countries such as Japan have even reached the stage of no longer being considered "ageing," but "aged" with 28 percent of its population projected to be over 65 by 2025. A more delicate choice of phrasing would be to regard the populations as "maturing" (rather like a good wine) rather than "ageing." The reasons for this change include (Coleman, 1993):

- Falling fertility rates—families are typically waiting until later in life to have children and are having fewer of them. Gone are the families with fourteen or fifteen children of two generations ago. Even the nuclear family average of 2.4 looks to be optimistic for many countries.

- People are living longer—100 years ago, typical life expectancy was under 50 years of age. Now the populations of many European countries, for example, can look forward to an additional 30 years of active life once they reach that age. In the US, the life expectancy at birth has risen from 47 years in 1900 to 76 years in 2000 (PBS, 2005). Medical advances, improvements in working conditions and the quality of our home environment (heating, clean water, etc.) look likely to continue to improve our ability to extend the number of healthy years that each generation can expect to live.

- Falling infant mortality rates—until recently the first five years or so of life were very high risk for many infants. Improvements in living conditions, medication and so on have reduced the infant mortality rates from 165 per 1,000 births in 1900 to just 7 in 2000 (PBS, 2005).

The combined effects of these changes on the global population are staggering. In a report published in 2002, the Population Division of the United Nations' Department of Economic and Social Affairs reported that:

> "In 1950, there were 205 million persons aged 60 or over throughout the world . . . At that time, only 3 countries had more than 10 million people 60 or older: China (42 million), India (20 million), and the United States of America (20 million). Fifty years later, the number of persons aged 60 or over increased about three times to 606 million. In 2000, the number of countries with more than 10 million people aged 60 or over increased to 12, including 5 with more than 20 million older people: China (129 million), India (77 million), the United States of America (46 million), Japan (30 million) and the Russian Federation (27 million). Over the first half of the current century, the global population 60 or over is projected to expand by more than three times to reach nearly 2 billion in 2050. . . . By then, 33 countries are expected to have more than 10 million people 60 or over, including 5 countries with more than 50 million older people: China (437 million), India (324 million), the United States of America (107 million), Indonesia (70 million) and Brazil (58 million)."

(UN, 2002)

Furthermore, the report concludes that:

> "The older population is growing faster than the total population in practically all regions of the world—and the difference in growth rates is increasing."

(UN, 2002)

The effects and implications of the ageing, or maturing, population are immense. For example, assuming that retirement age remains constant, the ratio of working age adults to retired adults in the UK is predicted to fall from 4.1 in 2003 to just 2.1 in 2045. In effect, approximately every two working age adults will have to pay enough tax to support one person's state pension (Goyder, 2005). In the United States, the post-war baby-boomer generation is due to start retiring in 2010, increasing the number of pensioners dramatically. This will place a huge burden on the Social Security system, which is predicted to become insolvent by 2040 unless tax rates are hiked upwards or the age of retirement is raised yet further beyond 67 years.

However, there is a potential source of hope for the future: productivity. There is an argument that what matters is not the ratio of workers to retirees, but workers to non-workers (otherwise known as the "economically inactive"), of which retirees are just one group. If that ratio can be decreased, then the effects of the increasing number of retirees can, to a large extent, be mitigated. Thus it is in the nation's best interests that as many people be offered employment as possible. Designing for accessibility can play a role in this. Potential workers who are currently unemployed because of their impairments could be brought into the workforce through suitably accessible workplaces—not just buildings and offices, but also employment tools, such as computers and telephones, etc. Another group is people who would otherwise have to be offered premature medical retirement because of conditions related to the ageing process, such as someone with developing arthritis, many of whom do not want to retire. Accessible workplaces can help defer medical retirement.

The costs of premature medical retirement are difficult to evaluate, but the Royal Mail in the UK conducted what is believed to be the first survey of its kind to estimate the cost to its business. It found that preventable premature medical retirement was costing the company over $200 million per year, without taking into consideration the costs of recruiting replacement staff or the loss of organizational memory (the knowledge and skills accumulated over the years by experienced employees). At the time, the company was losing $800 million per year and so the costs of preventable premature medical retirement became a major target for cost reductions. In case you were wondering, the definition of "preventable" premature medical retirement was where an employee was deemed capable of doing *a job*, just not one of *the jobs* available within the Royal Mail's working environment (Coy, 2002).

Implicit in the argument that the ageing of the population is relevant to design for accessibility is the fact that the ageing process is associated with certain decreases in user capabilities. In many cases, these are fairly minor losses of an individual's capabilities, but the minor losses can often have a cumulative effect. Thus someone whose eyesight is not quite as sharp as it was, whose keeps needing to turn the volume up a little bit louder on the television and whose fingers are not quite as nimble as they once were, may find some products as difficult to use as someone with a single, but more severe, impairment.

"1 IN 6 ADULTS HAS A DISABILITY"

Estimates of the prevalence of disability derived from any study depend on the purpose of the study and the methods used (Martin, Meltzer, and Elliot, 1988). Since disability has no "scientific" or commonly agreed upon definition (Pfeiffer, 2002), a major problem lies in the confusion over terminology. However, the International Classification of Impairments, Disabilities and Handicaps (ICIDH—WHO, 2001) represents a rationalization of the terminology frequently used. The ICIDH identifies impairment, disability, and handicap as consequences of diseases and presents a classification for each. This model can be extended to accommodate the effects of aging and accident—see Figure 2–1.

The ICIDH also defines disability as:

> *"any restriction or lack (resulting from an impairment) of ability to perform an activity in the manner or within the range considered normal for a human being."*

(WHO, 2001)

This definition has been used widely for both disability research (Martin, Meltzer, and Elliot, 1988; Grundy et al., 1999) and design research (Pirkl, 1994). However, such language is now generally considered too negative and it is preferable to describe users in terms of their capabilities rather than disabilities. Thus "capability" describes a continuum from high, i.e., "able-bodied," to low representing those that are severely "disabled." Data that describes such continua provide the means to define the populations that can use given products, thus leading to the possibility of evaluating metrics for a product's accessibility.

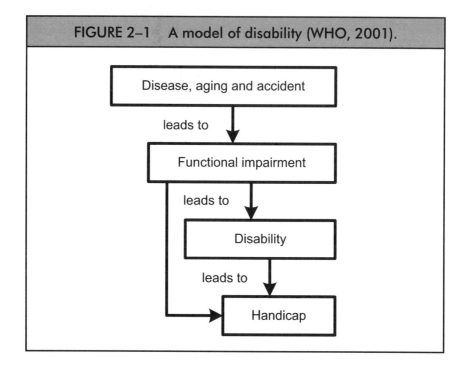

FIGURE 2–1   A model of disability (WHO, 2001).

Disease, aging and accident

leads to

Functional impairment

leads to

Disability

leads to

Handicap

American Community Survey

Accurate national statistics of the prevalence of impairments are difficult to locate, but fortunately not impossible. For example, the U.S. Census Bureau's 1999–2004 American Community Survey (USCB, 2004) asked respondents if they had any kind of disability, defined here as "*a long-lasting sensory, physical, mental or emotional condition*" (ACSO, 2006). Table 2–1 shows the prevalence of disabilities in the U.S. adult (16+) population.

In the UK, the Office of National Statistics has records of two detailed surveys that were performed in the late 1980s and 1990s that offer a detailed insight into the prevalence and level of severity of impairments across the entire British population (i.e., the combined populations of England, Scotland and Wales, but not Northern Ireland). The surveys involved over 7,500 respondents, sampled to provide representative coverage of the population.

The Survey of Disability in Great Britain

The Survey of Disability in Great Britain (Martin, Meltzer, and Elliot, 1988) was carried out between 1985 and 1988. It aimed to provide up-to-date information about

TABLE 2–1
The Prevalence of Disabilities in the U.S. Adult (16+) Population (After USCB, 2004)

Respondents: Population 16 years +:	Percent of Total 220,073,798	Margin of Error ±129,242
With any disability	16.0	±0.1
With a sensory disability i.e., with "blindness, deafness, severe vision, or hearing impairment"	4.7	±0.1
With a physical disability i.e., with a condition that substantially limits "walking, climbing stairs, reaching, lifting, or carrying"	10.6	±0.1
With a mental disability i.e., with a condition that makes it difficult to "learn, remember, or concentrate"	5.2	±0.1
With a self-care disability i.e., with a condition that makes it difficult to "dress, bathe, or get around inside the home"	3.1	±0.1
With a go-outside-home disability i.e., with a condition that makes it difficult to "go outside the home alone to a shop or doctor's office"	4.9	±0.1
With an employment disability* i.e., with a condition that makes it difficult to "work at a job or business"	5.6	±0.1

* Note—data for employment disability collected for ages 16–64 years only

the number of disabled people in Britain with different levels of severity of functional impairment and their domestic circumstances. The survey used 13 different types of disabilities based on those identified in the ICIDH (WHO, 2001) and gave estimates of the prevalence of each type. It showed that musculo-skeletal complaints, most notably arthritis, were the most commonly cited causes of disability among adults.

An innovative feature of the survey was the construction of an overall measure of severity of disability, based on a consensus of assessments of specialists acting as "judges." In essence, the severity of all 13 types of disability is established and the three highest scores combined to give an overall score, from which people are allocated to one of ten overall severity categories.

The Disability Follow-up Survey

The Disability Follow-up (DFS) (Grundy et al., 1999) to the 1996/97 Family Resources Survey (Semmence et al., 1998) was designed to update information collected by the earlier Survey of Disability in Great Britain (Martin, Meltzer, and Elliot, 1988). For the purposes of designing for accessibility, 7 of the 13 separate capabilities proposed in the Family Resources Survey and used in the DFS are of particular relevance, specifically:

- locomotion;
- reaching and stretching;
- dexterity;
- seeing;
- hearing;
- communication; and,
- intellectual functioning.

In the survey results, each capability is represented by a scale that runs from a minimum possible 0.5 to a maximum possible 13.0. Note though, that not all of the scales extend across this complete range, with some having maximum values of only 9.5. These individual capability scores may be grouped into three overall capabilities, computed via a weighted sum and mapped to a 0–10 scale utilizing the categories above:

- **motor**—locomotion, reaching and stretching, dexterity;
- **sensory**—seeing, hearing; and,
- **cognitive**—communication, intellectual functioning.

For example, a person's motion capability is derived from consideration of their locomotion, reaching and stretching, and dexterity, using a weighted sum as shown in Equation 2–1. The weighted sum is then mapped from the resulting 0–18.5 (the maximum possible upper limit) scale to a 0–10 scale.

$$\text{weighted sum} = \text{worst score} + 0.4 \times \text{2nd worst} + 0.3 \times \text{3rd worst} \qquad (2\text{--}1)$$

The survey results showed that a staggering estimated 8,582,200 adults in Great Britain (GB)—i.e., 20 percent of the adult population—had a disability according to the definitions used. Of these 34 percent had mild levels of impairment (categories 1–2—i.e., high capability), 45 percent had moderate impairment (categories 3–6—i.e., medium capability) and 21 percent had severe impairment (categories 7–10—i.e., low capability). It was also found that 48 percent of the disabled population was aged 65 or older and 29 percent was aged 75 years or more.

Multiple Capability Losses

Traditionally, designing for accessibility tends to focus on accommodating single, primarily major, capability losses. The reasons for this are two-fold. First, single major impairments are often the most noticeable and therefore are the easiest to inspire the necessary motivation to address them. Second, such impairments are the easiest to understand and are comparatively easy to compensate for, as there are no complex interactions with other capabilities.

Unfortunately, many people do not have solely single functional impairments, but several. This is especially true when considering older adults. Consequently, designers need to be aware of the prevalence of not only single, but also multiple capability losses. Therein lies a problem, as most user data focuses on single impairments.

Fortunately, both the American Community Survey and the Disability Follow-Up Survey provide valuable information for analyzing multiple capability losses. Tables 2–2 and 2–3 summarize the data extracted from those surveys. It is evident that in both surveys at least half of those respondents with some loss of capability have more than one loss of capability.

It is worth noting the comparative magnitudes of prevalence of motor and sensory impairments from the Disability Follow-up Survey (6.71 million and 3.98 million respectively). Many people automatically assume that designing for accessibility is really just designing for people who are blind. In practice, designing a product that is easier to handle and manipulate is likely to enable more people to use it than, say, adding Braille labeling. Only a comparatively small proportion of people who are blind have the skills to read Braille.

TABLE 2–2
Multiple Capability Losses as Reported in the 2004 American Community Survey
(After USCB, 2004)

Respondents: Population 5 years and over	Percent of Total 264,965,834	Margin of Error ±65,181
Without any disability	85.7	±0.1
With one type of disability	6.7	±0.1
With two or more types of disabilities	7.6	±0.1

TABLE 2–3
Multiple Capability Losses for Great Britain—Total Population ~46.9 Million Adults

Loss of Capability	Number of GB 16+ Population	Percentage of GB 16+ Population
Motor	6,710,000	14.3%
Sensory	3,979,000	8.5%
Cognitive	2,622,000	5.6%
Motor only	2,915,000	6.2%
Sensory only	771,000	1.6%
Cognitive only	431,000	0.9%
Motor and sensory only	1,819,000	3.9%
Sensory and cognitive only	213,000	0.5%
Cognitive and motor only	801,000	1.7%
Motor, sensory, and cognitive	1,175,000	2.5%
Motor, sensory, or cognitive	8,126,000	17.3%

Blindness is an important impairment that should not be overlooked when designing for accessibility, but has comparatively low prevalence. Even within the sensory impairment category, 1.93 million people have vision impairments compared with 2.9 million with hearing impairments. Note that 1.93 million + 2.9 million does not equal the 3.98 million in Table 2–3 because approximately 1 million people have both some hearing and some vision impairment, especially among older adults, and so the 3.98 million figure has been corrected to remove such double-counting. Of those 1.93 million people with vision impairments, the vast majority have low vision, i.e., they can see to some extent, but have difficulty reading regular size print, even with spectacles. Only a small percentage (less than 20 percent) of people with a vision impairment are classified as blind. Thus people who are blind constitute approximately only 5 percent of the total disabled population within Great Britain.

Effects of Impairment

Having looked at the prevalence of impairments, it is helpful to think about how each of those impairment types affects someone's ability to use a product. To illustrate how different users' capabilities can influence the difficulties that those users can expect to encounter, Table 2–4 shows the kinds of difficulties that users with specific impairments may encounter when trying to use a computer and the kinds of assistive technology that they may use. Table 2–4 represents broad difficulties and solutions. It is also instructive to think more deeply about how users with different impairments may be affected by common graphical user interface activities, for example clicking on an on-screen button:

> Consider a small button or icon on a software interface. Someone who is blind
> would not be able to see the button, or locate it. Someone with low vision may be

TABLE 2–4
Common Issues Facing Users With Functional Impairments When Trying To Use a Computer and Examples of the Assistive Technology That May Be Used (After IBM, 2005c)

	Vision
Issues:	■ Cannot use the mouse for input; ■ Cannot see the screen; or, ■ May need magnification and color contrast.
Assistive Technology:	■ Screen readers/voice output; ■ Braille displays; and, ■ Screen magnifiers.
	Hearing
Issues:	■ Cannot hear audio, video, system alerts, or alarms
Assistive Technology:	■ Closed captioning; ■ Transcripts; and, ■ "Show sounds".
	Motor
Issues:	■ Limited or no use of hands; and/or, ■ Limited range of movement, speed, and strength.
Assistive Technology:	■ Alternate input (e.g., voice); ■ Access keys; ■ Latches that are easy to reach and manipulate; ■ Single switches as alternatives to standard point and click devices; ■ On-screen keyboards; and, ■ Operating system (OS) based keyboard filtering.
	Cognitive/learning
Issues:	■ Difficulty reading and comprehending information; and/or, ■ Difficulty writing
Assistive Technology:	■ Spell checkers; ■ Word prediction aids; and, ■ Reading/writing comprehension aids.

able to see that there is a button, but not be able to recognize the type of button or read the text on it. Someone with a motor impairment may be able to see it and recognize it, but may not be able to position the mouse pointer or keyboard focus over it. Someone with a cognitive impairment or learning disability may be able to see, recognize (note—assuming that it is not described solely by words and that the learning disability is not a literary one) and activate the button, but may not know what it does. Next, consider a small button on a piece of hardware. Many of the same issues apply. Blind users may not be able to see or locate the button; low vision users may not be able to recognize it; motor impaired users may not be able to activate it; and cognitive impaired users may not know what it does.

Making the Business Case for Accessibility

"ACCESSIBLE DESIGN IS GOOD DESIGN"

One of the most persuasive arguments in favor of designing for accessibility is that an accessible design is often a good design. It is important to note that this proposition is *not* solely a direct result of a product being accessible—it is a result of the methods and practices that lead to the product becoming accessible. As we will examine in a later chapter, designing for accessibility builds on well-established user-centered design principles and usability methods. Those principles and methods have been developed specifically to create products that meet the needs of the users in terms of functionality, ease of use (Nielsen's practical acceptability—Nielsen, 1993) and also social acceptability. Thus any product that has been designed following such user-centered practices should, by default, be a well designed and, consequently, "good" product.

A common variant of this argument is the notion that a product that is designed for the users with the lowest levels of functional capability (i.e., the most severely impaired) is one that should be better and more accessible for everyone else. The logic behind this argument is easy to see. For example, while many people may have the functional capability to read small print in contracts, it would be much easier for everyone to read if it was not-so-small print, say 12, 14, or even 16 point fonts instead of 8 point. Similarly designing a computer program that includes an audio playback version of any written text not only benefits people who are blind, but also those who have poor reading skills (for whatever reason—be it dyslexia, illiteracy, or simply because the language used is not their mother tongue). Schematically, this argument can be thought of as advocating a top-down approach to design. Imagine the whole population is represented by a triangle—see Figure 2–2.

The vertical axis of the triangle represents increasing level of impairment, or reducing level of capability. Thus, at the top of the triangle are the people with the

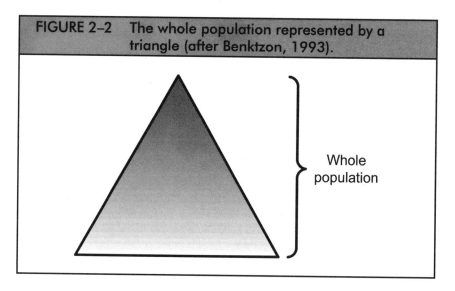

FIGURE 2–2   The whole population represented by a triangle (after Benktzon, 1993).

Whole population

most severe impairments and lowest levels of functional capability. At the bottom are all the fit, healthy people with no discernible impairments.

The logic continues that if a product is designed for users with the lowest level of capability (the top of the triangle), then "gravity" means that the product will permeate its way down the triangle and be acceptable to all the people below—see Figure 2–3.

While this is a conceptual straightforward and somewhat appealing argument, in practice there are some holes in it. For example, designing a product that has Braille on it only benefits a subset of blind people (as discussed earlier, only a comparatively small proportion of people who are blind have the skills to read Braille) and thus would appear to be of limited use to someone who does not read it. Similarly, including support for hearing aids only really benefits people who use them. However, the presence of these aids is invaluable to those who need them and are unlikely to detract from the user experience for people who do not.

A more difficult case to resolve is, for example, the design of a mobile telephone. Most potential users of these telephones are typically interested in a product that is as small and light as possible. This, in turn, leads to small buttons (not good for someone with a motor impairment) and even smaller text labels on the buttons (not good for someone with low vision). Designing a mobile telephone for those potential users would require larger buttons that are easier to press with larger text labels. However, these requirements conflict with the principal design goal of a telephone that is small and light.

Thus, the conceptually and ideologically attractive argument in favor of top-down approaches is somewhat limited in its pragmatism. A more pragmatic approach for companies, albeit one that is less ideologically "pure" is to take a bottom-up approach—as illustrated in Figure 2–4.

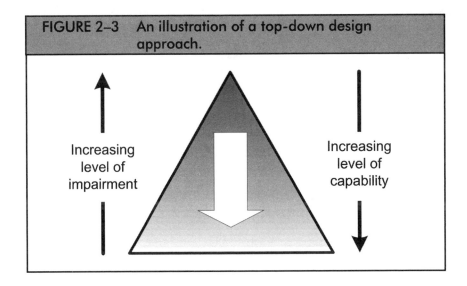

FIGURE 2–3   An illustration of a top-down design approach.

Increasing level of impairment

Increasing level of capability

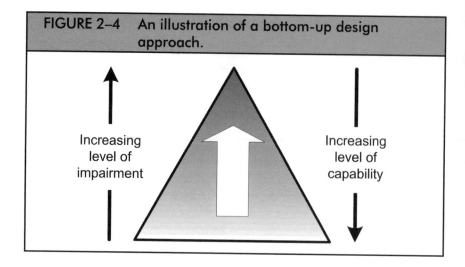

FIGURE 2–4   An illustration of a bottom-up design approach.

Increasing level of impairment

Increasing level of capability

In the bottom-up approach, a mainstream product is re-designed to reduce its demands on the user and thus move the product further "up" the population triangle. This approach is attractive to many companies because it involves an evolution of existing products rather than the larger-scale and potentially risky revolutionary top-down approach.

However, this approach also has its limitations. By basing much of the approach on existing mainstream technology, there is a danger that if the existing technology is simply too limited in its ability to be adapted to the needs of users with more severe impairments or capability losses, then the potential ability to move up the triangle is also limited. Designs can often only evolve so far before a degree of revolution is required. For example, a mobile telephone can only be modified so much (e.g., buttons only made *this* big) before a radical re-think of its form is required to support larger buttons and so on.

The concept of countering design exclusion is based on a marriage of both top-down and bottom-up approaches. It takes the pragmatic bottom-up approach as a starting point—basically find which features and functions of the product cause an unnecessary functional capability demand on the user and re-design them to lessen or remove that demand. At the same time, elements of a top-down approach can be added. For example, specific solutions, such as adding voice output or hearing aid support, should be added wherever possible and appropriate. This idea of marrying the top-down and bottom-up approaches underlies much of the theoretical basis of this book.

"RELYING ON THIRD-PARTY TECHNOLOGIES IS BAD BUSINESS"

Designing a successful product that is accessible requires that accessibility be considered as an integral component of the product and its design process. Many companies do not want to invest time and effort into making a product accessible, because they believe that their customers can purchase third-party assistive technologies to help them access that product; so why should they have to worry themselves with making their product accessible? This is half-true. Assistive technologies are available for all kinds of purposes and can be remarkably useful for those who need to use them. However, they are also invariably quite expensive to purchase and also often have to be adjusted to each user and each product. This process can be time-consuming and also add to the overall cost of ownership of the product for the user, especially if technical assistance is required to perform the adjustments.

Examples of assistive technologies range from simple foam rubber grips that fit over handles to make them easier to grasp, to fully integrated "smart home" technologies, with remote controlled window blinds, door locks, heating controls and so on. Spectacles and hearing aids are probably the most commonly encountered assistive technologies.

There is a significant subset of assistive technologies focused on enabling access to computers. Input devices can be modified through the use of add-ons such as keyguards (a sheet of plastic that fits over the top of a standard keyboard and reduces the likelihood of two keys being pressed simultaneously), that are used to modify a standard mouse or keyboard. Alternatively specialist devices can be used, such as oversize trackballs or voice recognition systems. The user interface can be tweaked using options built into the operating system, for example most offer on-screen keyboard emulators, magnification options, etc. Finally, screen-readers can be used to change the output from being visual in nature to aural, by literally reading out whatever text is displayed on-screen.

The reason why many assistive technology solutions are expensive is that they are often developed for a very particular niche (and thus limited) market while at the same time having quite high development costs. The development costs are so high because many of these products are purchased by residential care centers and hospitals, thus they often fall under the category of medical devices, with all of the concomitant testing that involves. Even if they do not classify as medical equipment, they do still need considerable design and development effort, because their target user population is typically severely impaired and, thus, the most challenging population for which to design. Consequently, designing a product that relies on the user purchasing some form of assistive technology to operate it will automatically make that product significantly more expensive for the user.

Additionally, designing a product that depends on another company's products (the assistive technology) is exposing the product developers to risks beyond their control. What happens if the assistive technology company goes bankrupt and the assistive technology is no longer available? What happens if the third party changes the design specification and that assistive technology no longer works with the product? Screen readers, for example, often operate by using features within an operating system that are, technically speaking, unsupported by the operating system provider. Thus, whenever the operating system is upgraded, there is a distinct possibility that the screen reader will not work on the new version until it is also upgraded. That upgrade may force a change in behavior on the screen reader (if the operating system did something one way in the previous version, but another way in the new one), which, in turn, may mean that it does not operate with the product any longer. Any company that willfully leaves itself subject to these vagaries of chance when it has the means to re-design its products not to need to rely on other companies' products is embarking on a high-risk business strategy.

Another argument against relying on the use of assistive technologies is that, by definition, the assistive technologies are retrospective adaptations to the product and constitute an attempt to fix or overcome a feature or function that is inaccessible to certain users in its unmodified form. As such, the resultant combination of the product and the assistive technology is extremely unlikely to offer a seamless and smooth interaction as there are now two interfaces and interaction models to consider—that of the product and that of the assistive technology. Instead, the interaction is much more likely to be clunky, awkward and complicated, and consequently much more likely to be unacceptable to those users.

"ACCESSIBILITY BEST PRACTICES ARE BECOMING READILY AVAILABLE"

One of the key anxieties or concerns that a company may have when considering design for accessibility is that it may appear to represent a significant risk. It may appear to be a completely new design doctrine, is undoubtedly a challenge to do well and even if the company invests heavily in this area, it may not necessarily see a return on its investment. These concerns are primarily a variation on the typical fear of the unknown. The best way of overcoming them is to identify well-trodden paths that other companies have followed and seen success with. Some companies, such as IBM and Microsoft, have accessibility resource centers that distribute detailed information on designing for accessibility. These are excellent resources, although they are likely to be biased to represent the company in the best possible light.

Alternative resources are available from not-for-profit organizations. Many charities offer consultancy services to companies and often have examples of successful product designs on their Web sites (see, for example, the Royal National Institute for the Blind Web site).[1] The Design Council also hosts an excellent web resource with an extensive set of successful case study products (Coleman, 2005).

Furthermore, standards bodies are beginning to encode many of the best practices from such case studies into reference materials. Perhaps one of the best-known examples of these is from the Web Accessibility Initiative (WAI) of the Worldwide Web Consortium (W3C). The WAI working group has developed a set of Web Content Authoring Guidelines (WCAG) that list a set of design objectives that

---

1. http://www.rnib.org.uk

should be met when designing Web sites to be accessible. Furthermore, these are accompanied by a set of checkpoints that have to be met to validate the accessibility of the site. The checkpoints are subdivided into three priority levels (W3C, 2005):

- **Priority 1—most important**—failure to satisfy this level means that one or more user groups will find it *impossible* to access the site;
- **Priority 2—important**—failure to satisfy this level means that one or more user groups will find it *difficult* to access the site; and,
- **Priority 3—desirable**—failure to satisfy this level means that one or more user groups will find it *somewhat difficult* to access the site;

The WAI guidelines are not without their critics, most notably for the occasions when the guidelines overstep the mark into suggesting what a Web site's content should include rather than just providing recommendations on how the site should be implemented and coded. Additionally, the guidelines are also heavily biased towards blindness and not so much towards other disabilities. However, they make a very good template for how design guidance can be structured for designing for accessibility.

"ACCESSIBILITY IS BECOMING A HOT TOPIC"

Accessibility is becoming a hot topic of which businesses have to be aware. For example, when the Commonwealth of Massachusetts declared in 2005 that it wanted to adopt the Open Document format (ODF), it was the disability representatives that were most active in challenging this decision. The only other party that was comparably active was Microsoft, which stood to lose its dominance in the office application market in the Commonwealth (Microsoft currently does not support ODF in its Office products, preferring to support its proprietary Office Open XML format instead). The concern of the advocacy groups was that disabled employees within the Commonwealth would lose the ability to use the assistive technologies (designed to work with the Microsoft Office applications) that they were used to using for access to Commonwealth forms and publications.

The result of this sudden highlighting of accessibility concerns was (at the time of writing) that ODF is being recast to make it more accessible. In the meantime, the Commonwealth has declared that anyone who cannot access the ODF versions of its documents and forms will be able to use whatever software is accessible to them. Thus a decision that was taken with the very best of intentions, i.e., to make sure that the Commonwealth was not dependent on one supplier's proprietary document format for all of its official records, documents and forms, was potentially jeopardized by not taking a sufficient account of accessibility. It is arguable that many of the disability advocates' claims were based on a questionable interpretation of the Commonwealth's decision, but the end result is not in question. ODF is becoming more accessible because of the issues raised by the advocacy groups.

SUMMARY

In this chapter we have seen that there are several arguments that can be put forward when asked to provide a business case in favor of designing for accessibility. The ageing population and the approximate 20 percent prevalence of disability in the adult population demonstrate that there is a significant number of people who would benefit from more accessible products. This does not even include people who are impaired by their environment (e.g., on a moving train). The argument that accessible design is good design will be illustrated throughout the rest of this book.

The emergence of design best practices, reinforced through standards, is probably one of the most useful developments in the field of designing for accessibility. These practices and standards give designers and companies a tried-and-tested structure for them to follow that they can have confidence in and also will help them avoid the worst pitfalls.

KEY POINTS

- Accessibility represents a business opportunity.
- Accessibility challenges designers—and the best designers respond to that challenge.
- Accessibility needs to be an integral component of the product—and considered right from the very outset of the design process.

3

Implementing Design
for Accessibility
in Industry

Companies are increasingly finding themselves having to ensure that their products and services are accessible and inclusive, or else be exposed to the possibility of litigation and damage to their brand reputation. However, the adoption of inclusive design within industry has been patchy at best. While there are undoubtedly companies that have yet to be persuaded of the merits of inclusive design, there is a growing number that want to design inclusively, but do not know how to set about doing so (Keates and Clarkson, 2003).

As discussed earlier, disability is not a simple consequence of a person's impaired capability, but results from a failure to take proper account of the needs and capabilities of all potential users when designing products and services. Internationally there is a move towards creating a more inclusive society. This move is being encouraged through a mixture of legislation, regulation, incentives and also by changing attitudes within society. Research has shown that despite the various incentives for designing for accessibility, many companies have not done so. Prevalent among the barriers cited by those companies are:

- the perception that designing for accessibility is primarily a niche activity that does not apply to them; and,
- concerns about the knowledge and skills requirements involved.

Designing for accessibility is not simply an ethical goal, but can be highly beneficial for a company if implemented correctly. on the other hand, companies that do not adopt such an approach when developing new products will find themselves at risk of punitive legal actions and also losing ground to more inclusive and accessible products and services from competitors.

RATIONALE UNDERLYING DESIGN FOR ACCESSIBILITY

It is not always possible or economically viable to design one-product-for-all. Thus the rationale we shall use here is that of countering design exclusion, i.e., removing as many barriers to use as possible that have arisen as a result of design decisions made. This rationale advocates the systematic identification of the capability demands placed upon a user by a product or service. Where those demands exceed the capabilities of the target users, then the feature responsible for the demands should be re-designed, wherever possible, to remove the cause of the exclusion.

For example, consider a newspaper that is printed using a very small font. A proportion of the potential readership will be unable to read the newspaper because the vision capability demanded of them by the small font is greater than they possess. In other words, their eyesight is simply not good enough to read such a small font. Thus, the choice of font size has excluded them. The solution is therefore straight-forward-to increase the font size. Doing so would remove the source of exclusion and include many more users. The newspaper will have become more inclusive. However, if the font size is continually increased, then the newspaper will eventually become bigger and bulkier, thus making it more expensive to produce (more paper needed) and also potentially more exclusionary as the readers require greater dexterity skills to manipulate the larger and more numerous pages. This leads to the inclusion of the caveat "as reasonably possible" in all of the guidance provided in the next couple of chapters.

Note that while the name "design for accessibility" implies that it is a single entity or process, in practice there are numerous methods, tools and techniques that can be used to achieve the goal of inclusive design. Those methods, tools and techniques can be referred to collectively as *design for accessibility practices*. Another very important point to be aware of at this stage is that the term "users" does not just mean customers; it also applies to employees and also those involved throughout the supply chain. All design for accessibility best practices, from philosophy and audits to implementation of more accessible solutions, can be applied to the design of business processes and work environments every bit as much as to the product-line.

FIRST STEPS IN DESIGN FOR ACCESSIBILITY

This chapter, along with the next two chapters, provides an approach not only to the adoption of design for accessibility by companies, but also to the recognition that management has a key role to play in ensuring that design for accessibility becomes the norm, rather than the exception.

The Role of Management

There are three distinct levels of hierarchy within companies and that each has a different role to play in the adoption of designing for accessibility practices. At the top of the corporate hierarchy, are the individuals who comprise the senior management, who have the responsibility of initiating the adoption of inclusive design practices across the company. They have the power to drive this transformation in company outlook and practices. Suggested activities for senior management to drive the adoption of design for accessibility are presented in Chapter 4.

Below the senior management is the middle management. For our purposes here, this level consists primarily of executives who are responsible for, and oversee, design and marketing activities. Middle management plays more of a shepherding role in the adoption of design for accessibility. They are responsible for ensuring that corporate policies and strategies are implemented as envisioned and that there is no unnecessary deviation. In the event of conflicting design requirements, middle managers are the first line of decision-making over what the compromise solution should be. The role and responsibilities of middle management are discussed in Chapter 5.

Central to the approach advocated here is the concept that design for accessibility is not simply a bolt-on activity that is the responsibility of the designers. Instead, it needs to become a corporate ethos, with a company-wide culture promoting and encouraging the adoption of design for accessibility practices. It is commonly accepted that those who manage design tend to have a much greater impact on the outcome of projects than those at lower levels of the company hierarchy. In other words, success in adopting design for accessibility is achieved as much from the top-down through the company hierarchy as it is from the bottom-up. Companies adhering to the approach provided in this book are encouraged to take steps such as developing corporate inclusive design mission statements and appointing a principal as the inclusive design champion.

The Pervasive Nature of Design for Accessibility

Companies should consider solutions across the entire brand and not just on a product-by-product basis. for example, companies could try to develop all of their products to be as inclusive as possible (along the lines of one product for all). Alternatively, they could consider:

1. adding more accessible products to their existing ranges;
2. designing add-ons to improvement the accessibility of their current products; or,
3. developing completely new ranges of products.

It is also worth noting that the term "product" is used for brevity in the next few chapters in accordance with terminology used in Quality Management Standards (ISO9000:2000 series), to refer to products, services, processes (including business processes), environments, and interfaces.

The Role of Written Records

It is essential to keep written records of all decisions made that affect the final accessibility of the product. The aim is to develop a paper trail so that in the case that a company is subject to litigation, documentary evidence of the rationales behind the decisions is easy to obtain. The presence of the paper trail also encourages companies to ensure that users are only excluded where there is a real, substantive and justifiable reason that they could defend in court, otherwise their own paper trail could condemn them as easily as it could exonerate them. It is in their interests to strive for their products to be as accessible as possible.

Although this advocacy of keeping written records implies that the approach provided here is very bureaucratic in nature, it is not. The aim is to recognize and promote the vision of inclusive design and that it is better to lead through embracing the challenges of inclusive design in a positive way. It supports the assertion that when designers are inspired with the new perspectives that inclusive design offers, they often respond with leaps forward in thinking that raise the chances of genuine innovation. That innovation often results in competitive advantage and, hence, greater profitability.

BS7000-6—GUIDE TO MANAGING INCLUSIVE DESIGN

The steps outlined in the following two chapters represent the basic activities that companies should engage in when designing for accessibility. For companies looking for a complete point-by-point framework advocated by a standards body, a new technical guidance standard has been published by the British Standards Institution (BSI)—*BS7000 Part 6: 2005—Guide to managing inclusive design*, (henceforth referred to as BS7000-6)—as part of the BS7000 Design Management Systems series.

The emphasis on the management of inclusive design makes this new standard unique among the many standards that address accessibility issues. Prior to BS7000-6, most standards focused exclusively on specific topics such as:

- buildings access—e.g., BS 8300:2001—Design of buildings and their approaches to meet the needs of disabled people. Code of practice (BSI, 2001);
- assistive technology—e.g., ISO 9999:2002—Technical aids for persons with disabilities—Classification and terminology (ISO, 2002);
- anthropometric measures—e.g., BS 4467:1997—Guide to dimensions in designing for elderly people (BSI, 1997); and, even,
- standards development—e.g., PD ISO/IEC Guide 71:2001—Guidelines for standards developers to address the needs of older persons and persons with disabilities (ISO, 2001).

As with most British Standards, BS7000-6 was written by a drafting committee, of which I was a member. The standard is a guide that puts forward a framework of best practice. As such, no organization is forced to comply with it, though compliance may be specified in a contract between organizations. It is a valuable information resource as it provides probably the most comprehensive guide available on how a company can change both its culture and processes to become more inclusive.

Note that while the focus of this book is on design for accessibility, BS7000-6 focuses on inclusive design. Designing for accessibility is an important facet of inclusive design. As was discussed in the opening chapter of this book, inclusive design encompasses designing for users based on social and cultural factors, as well as physical and cognitive capabilities. However, while all of the recommendations included in this book are based on designing for accessibility, they are exactly the same steps that should be considered when adopting inclusive design.

BS7000-6 may also play a potentially pivotal role in any possible litigation suits against a company. As discussed earlier in this book, the majority of existing legislation requires "reasonable accommodations" to be made. Since the concept of "reasonableness" has yet to be adequately defined in various courtrooms around the world, all companies would be well advised to always have one eye on the possibility of being on the wrong end of a lawsuit. Under such circumstances, anything that demonstrates a willingness to avoid discrimination between customers or employees will be valuable in the company's defense. Therefore, voluntarily implementing a standard such as BS7000-6 has three potential benefits for a company:

> First, it should reduce the chance of litigation, since all products and workplaces will have been carefully examined for potential exclusion. Second, it encourages companies to maintain a clear and simple paper trail of how decisions about the level of inclusivity (and, consequently, exclusion) were reached. Such a paper trail is often invaluable in a lawsuit. Finally, should a company be sued, the fact that it has demonstrably at least made the effort to embrace inclusive design practices is likely to count in its favor—although whether that is sufficient to defend a case depends on the merits of the particular complaint.

> *It is important to underline that compliance with a standard will never confer complete protection from prosecution.* Standards represent a consensus of accepted and proven best practices available at the time of writing. Unless updated regularly, there is always the opportunity that they can be overtaken by new developments. Consequently, while compliance with BS7000-6 is an excellent first step, the onus still remains on a company to keep re-visiting how accessible their products and workplaces are.

SUMMARY

Implementing design for accessibility within a company requires planning and forethought. Its implementation is, in effect, a business transformation process, realigning a company's product ranges to meet the needs of the users. In doing so, this offers a significant opportunity for realizing genuine innovation. The following chapters describe the basic approach required to adopt design for accessibility on a company-wide basis.

KEY POINTS

- Design for accessibility is important for industry.
- Adopting design for accessibility practices requires a plan. It does not just happen overnight; it needs a strategic approach.
- Design for accessibility affects all levels of the corporate hierarchy within a company.

4

The Role of Senior Management

At the top of any company structure is the senior management. The ultimate responsibility for the adoption of design for accessibility rests squarely on their shoulders. They should ensure that an appropriate culture of design for accessibility is adopted throughout the company. They are also responsible for the creation of a paper trail documenting how the company is responding to the challenge of designing more accessible products.

As shown in Table 4–1, the suggested activities for a company's senior management are divided into five main phases and each phase is comprised of explicit stages. Each stage needs to be fully documented and approved by the person or committee designated responsible for meeting accessibility targets before progress to the next stage is possible. Also remember, that it is highly unlikely that a perfect plan will be arrived at and implemented on the first pass through these tasks. An iterative approach will most likely be required, with frequent re-visits to the preceding stage and it may even prove necessary to re-visit even earlier phases and stages.

When reading through the following phases and stages, it may be helpful to keep the analogy of designing for safety or for quality in mind. As discussed earlier in this book, design for accessibility does not lend itself to being a bolt-on activity. It *can* be bolted on, but the results are usually less than satisfactory. Much better is to view it as an integral activity, present right from the outset of each product development cycle.

Both this chapter and Chapter 5 provide basic frameworks of issues that organizations should address when looking to adopt design for accessibility. It is intended to help most companies begin their adoption of design for accessibility practices and is more than sufficient for that purpose. This is a launching point for the design for accessibility journey with helpful waypoints marked out. It is not a fully mapped-out route, as that route will vary for each company.

TABLE 4–1
The Design for Accessibility Implementation Framework for Senior Management

PHASE 1—Scoping the Business Plan
Stage 1.1: Assigning responsibility
Stage 1.2: Acquiring basic knowledge
Stage 1.3: Understanding the current situation
Stage 1.4: Formulating a plan of action

PHASE 2—Shaping the Business Plan
Stage 2.1: Communicate design for accessibility intent
Stage 2.2: Define corporate philosophy
Stage 2.3: Identify specific objectives to be achieved
Stage 2.4: Promote design for accessibility across the company

PHASE 3—Implementing the Business Plan
Stage 3.1: Implement management structures for design for accessibility
Stage 3.2: Perform pilot studies
Stage 3.3: Recognize and enhance expertise
Stage 3.4: Review progress

PHASE 4—Selling Accessibility
Stage 4.1 Identify and leverage competitive advantages
Stage 4.2: Identify opportunities for improved corporate image

PHASE 5—Reviewing and Refining Business Plan
Stage 5.1: Recognize and reward successes
Stage 5.2: Review and refine design for accessibility approach

PHASE 1—SCOPING THE BUSINESS PLAN

The first phase of any business transformation process is to develop an understanding of the scope of the transformation, its magnitude and what is involved.

Stage 1.1—Assigning Responsibility

The *ultimate* responsibility for ensuring design for accessibility practices are followed rests squarely on the shoulders of the principals and, in particular, the chief executive of the company. However, other than in very small companies, that person cannot bring about the necessary corporate changes single-handedly. An appropriate system of delegation needs to be implemented, but with the chain-of-command (and of communication, responsibility and accountability) clearly pointing back to the chief executive.

Suggested outcome—*a named top manager to champion and have explicit responsibility for design for accessibility.*

Stage 1.2—Acquiring Basic Knowledge

A company cannot just begin designing accessible products without first achieving a basic level of understanding of what is involved in implementing design for accessibility. The first step, especially for a company moving into this area for the first time, is to become familiar with the subject area. This initially involves high-level training to learn the terminology, basic principles and so on. It also involves identifying appropriate sources of information to contact. These sources can range from prominent individuals in the area, to organizations dedicated to providing information on best practices. Many of these issues are addressed in the chapter on *Filling the Skills Gap* (Chapter 6).

Suggested outcome—*a common understanding of the basic aims and principles of design for accessibility.*

Stage 1.3—Understanding the Current Situation

When embarking on a new change of direction, it is often helpful to begin with small steps. Audits are an excellent technique for highlighting the initial areas to be focused on and, in this case, should focus on two objectives. For a company adopting design for accessibility practices for the first time, a good first stage of the audit process is to identify what the required levels of accessibility are. Another way of thinking about this is to view it as a risk assessment exercise. With the proliferation of anti-discrimination legislation, for example, a company would be well-advised to determine whether its current product offerings (and workplace, for that matter) leave it open to the risk of lawsuits.

Establishing this involves identifying any particular legal or standards requirements affecting a particular type of product or the workplace as a whole. Thus, for example, an IT company based in the United States will most likely need to identify the accessibility requirements of Section 508 (WIA, 1998). Alternatively, a large company in the UK will need to establish what their required quota of employees with disabilities is. For a more experienced company, these objectives should be more rigorous and look to move beyond meeting the minimum levels required for compliance and focus more on overall quality of user experience.

Having identified what the target levels of inclusion (or exclusion, depending on which way you prefer to look at it) are, the company then needs to audit how well it is meeting those targets. For a company with an existing portfolio of products, which will apply to most companies, this typically involves reviewing how accessible and inclusive those products are. This can be achieved by using either absolute or relative metrics. Absolute metrics are those that place the products on a fixed scale. An example of this is calculating the level of design exclusion, i.e., how many people in the general population will experience difficulty when trying to use a particular product (Keates and Clarkson, 2003a). Here the scale used is either number of people, or proportion of the population. Another example of an absolute scale is the W3C Web Accessibility Initiative's checklists (W3C, 2005), where Web sites are rated on A, AA, or AAA compliance.

Relative scales are based, for example, on how the company's product range compares to those of their competitors. The aim is to answer questions such as "Are we ahead of the game? Behind it? Not even on the field?" This type of competitive analysis is common in usability and other areas, and would take only a small amount of modification to consider the level of exclusion (and its corollary, accessibility).

The audit should not only address the current levels of accessibility of the product range. It should also address *why* the current levels are what they are. Identifying particular design methods, marketing criteria, decision-making processes, etc., that encourage or lead to inaccessible products should be identified and examined as early as possible.

Suggested outcome—*a completed audit of the company's entire or selected product lines.*

Stage 1.4—Formulating a Plan of Action

Once the size and scope of the opportunity for transformation and innovation has been established, the company principals need to decide on a plan of action and implement it. At this stage, a basic statement of intent is probably satisfactory, but the more detail that can be decided up front, the easier the plan will be to put into place. Once the four stages of the first phase are complete and documented adequately, the senior management of the company should be ready to address the second phase.

Suggested outcome—*an initial plan of action for the implementation of design for accessibility.*

PHASE 2—SHAPING THE BUSINESS PLAN

Having established a plan of action in principle, the company needs to begin implementing it. The first stage of doing so is to establish a solid foundation on which to base the necessary corporate changes.

Stage 2.1—Communicate Design for Accessibility Intent

It is essential that everyone involved in the process of developing and delivering new products and services understands how important it is to not exclude potential customers from being able to use the company's products. Clear communication is central to this and the opportunity is there for company principals to take an unequivocal stand that speaks to all company employees. Since the message underlying design for accessibility is a fairly simple one, it lends itself to a corporate-wide mission statement. Mission statements are frequently maligned as a result of their over-use in areas where they are not well suited and being dumping grounds for excessive corporate-speak. In this case, the statement can be kept short and to the point. For example, as a starting point, try the following:

> "Our mission is to create products that are accessible to all of our customers, whether current or potential. We will not discriminate against customers on grounds of age, gender, expertise or functional capability."

A more sophisticated and ambitious mission statement would be:

> "Our mission is to identify and leverage the currently ignored potential in all of our products and services to meet the needs and aspirations all of our potential customers both in currently targeted segments and new sectors."

Note that in both of these mission statements, design for accessibility is not presented as the ultimate goal. Instead, it is positioned as a means to achieve greater customer satisfaction.

Suggested outcome—*a mission statement that clearly communicates how important design for accessibility is for the company.*

Stage 2.2—Define Corporate Philosophy

Having established a high-level statement of intent through the mission statement, it is time to begin adding some flesh to the bare bones of the company's approach to design for accessibility. The mission statement is a good starting basis for the overall design philosophy of the company. However, this philosophy needs to be reviewed, developed and refined at appropriate intervals as the company accumulates experience and expertise.

There are several comparatively straightforward philosophy issues that can be decided at this stage. For example, the company can choose the language that is to be used to describe its design for accessibility program. It can stay with "design for accessibility" or adopt "inclusive design," "universal design," "universal access," "design for all," and so on. Although each of these terms varies slightly in its details, the basic objective of each is the same, namely "countering design exclusion." However, the choice of language used needs to be based on the understanding that each option is associated with a particular set of preconceptions in people's minds. For example, universal design has a very specific set of defined attributes and is commonly associated, rightly or wrongly, with designing for people with disabilities (see the discussion on international approaches to designing for accessibility in Chapter 1). Thus a company should choose the approach that most closely reflects the company's other philosophies. A good example of how the right name can make a difference is that of office-supply giant Staples and its new slogan "that was easy"—a very simple premise that says everything it needs in three words.

Another philosophical decision that can be made at this point is whether the company will express a preference for bottom-up or top-down design approaches.

Will it look to remove unnecessary impediments to use from its existing mainstream products through the principles of countering design exclusion, or design wholly new products for the least able customers and try to market them to appeal to their mainstream customer base?

Suggested outcome—*a design for accessibility "bible" outlining corporate philosophy and preferred language.*

Stage 2.3—Identify Specific Objectives to Be Achieved

Looking at specific objectives that a company may set itself, it is useful to reflect back on the audit that was performed in Phase 1; an immediate corporate objective should be to reach an improved level of accessibility both across the product-range and within the workplace. Thus, for example, one target would be to reduce the number of potential customers who cannot use a product by 20 percent on a year-by-year basis. A more common target may be to match or beat the current market-leader. The *absolute minimum* target should be to meet all legal compliance targets.

A major objective at this early stage is how design for accessibility relates to the positioning of the company's branding. Recognizable brands are very marketable entities and are often the result of many years of careful planning and nurturing. Some brands are designed to appeal to younger, more active adults. Others are designed to appeal to families and so on. Many larger companies own multiple brands and brand names. Some brands, such as those that explicitly target older adults, lend themselves well to openly advertising the role that design for accessibility played in the development of products. Other brands, such as active sports brands, do not. Companies have to decide whether to modify their existing brand positioning, or possibly develop new brands in the light of their more design for accessibility practices. Some companies may decide not to tinker at all with the existing brands and not make any fuss at all about having designed a more accessible product and simply let the quality of the design speak for itself.

At this fairly early stage of adopting design for accessibility practices, the strategies for achieving these objectives are likely to be quite blunt instruments that will need honing over time and with increasing familiarity with the subject within the company. Perhaps the most useful strategy at this point is to begin identifying the skills available within the company and bringing people with the most relevant skills together into a recognized corporate-wide team. The purpose of the team would be to manage the change program, i.e., the process by which current company practices are modified to become more focus on design for accessibility. If chosen well, that team may even be able to generate enough impetus on its own to achieve many of the required objectives. However, in practice, it is highly likely that the explicit backing of the company's senior management may be necessary to effect the necessary changes across the company.

Whenever change is required, it is easier to implement through persuasion rather than coercion. Depending on the existing company culture, the realignment of company priorities to embrace design for accessibility will meet varying levels of acceptance (and resistance) in different parts of the company. Rather than take the proverbial bull-in-a-china-shop approach, a steady drip feed of information that builds into a torrent of enthusiasm, is a much more effective approach. This new realignment of company objectives should be communicated with due care through a proper communications plan that has been formulated to get the core messages across without overloading or scaring anyone unduly. Design for accessibility is the kind of activity that employees and management need to feel enthused about for it to be effective. Ideally, they should regard it as a challenge to be embraced, because they see the benefits and all will be equally stretched to achieve the goals set.

When it comes to implementing the change program what has to be avoided at all costs is the culture of the "checkbox police." In other words, if design for accessibility is reduced to simply another checkbox item to be ticked before a product can be shipped, then the company has not fully understood what design for accessibility is really about. Not only that, but such an approach lends itself to internal dissension within the company. The scenario under which this happens is alarmingly common. It goes as follows:

> A company appoints a team to be responsible for ensuring that products meet specific targets for product accessibility before they can be shipped. Meanwhile, the design team responsible for creating new products is being pressured by the marketing people to get a new widget ready for sale. At the same time, the production team is pressuring the designers not to create anything that will cause them headaches on the production line. The design team is vaguely aware of the need to meet accessibility targets, but has not really bought into the idea. So that team focuses on doing what it knows how to do, that is developing the widget to meet a functional specification drawn up by the market research team. The widget design process is ultimately completed and begins its final checks before being shipped. At this point the design for accessibility team receives the widget for approval. They discover that the widget does not comply with the company's stated accessibility targets. The team is then left with no alternative but to fail that item on the checkbox and send the widget back for redesign. The design team becomes extremely unhappy with this, as does the marketing team, and both begin to resent the "checkbox police" as the design for accessibility team becomes known, most likely behind their backs. The scene is set for rival tubthumping as the members of the design for accessibility team point out that they are just doing their job.

Looking at this scenario in these terms, the solutions to avoid these problems are quite clear. For a start, the initial design specification needs to spell out the design for accessibility requirements explicitly, so no one is in any doubt what the targets are. Second, members of the design for accessibility team should be integrated tightly into the design process, making their expertise available on-demand. Their first sight of a product should never be just before it is due to leave the factory. So, why is the image of the checkbox police so common?

The simple answer to this is other business pressures. A design team is typically operating to a very tight deadline. All that they often *really* care about is getting the product out of the door as quickly as possible. In an earlier book (Keates and Clarkson, 2003) I described this phenomenon as "the tyranny of time to market." Adding another set of design constraints and objectives to be met is about as welcome as, say, halving the design team's budget. Consequently, unless the design team *and its management* are motivated appropriately, they are likely simply to pay lip service to those requirements that they do not see as core to the product. This is only human nature, and they would argue that it represents effective time management on their part.

Should the phenomenon of the "checkbox police" begin to appear within the company, this is not necessarily the fault of the design for accessibility team's management—although it is entirely possible that they may be contributing to the situation. It is, however, primarily a symptom of design for accessibility not being central to the design process, which in turn is a consequence of design for accessibility not being viewed as a corporate priority. Thus senior management needs to ensure that everyone in the company is aware of how importantly design for accessibility is viewed.

Suggested outcome—*a master program of clearly stated initial corporate objectives with an identified time-line for completion.*

Stage 2.4—Promote Design for Accessibility Across the Company

Going back to the drip-feed idea, an interesting option for initially getting the workforce thinking about design for accessibility is a set of posters showing either users having difficulty with a product, or someone with a physical disability taking part in an everyday activity. The posters should not offer any easy answers—the aim is to be thought provoking. Rather like an extended advertising campaign on the television, the posters can be adapted over time to tell a much bigger story, ideally highlighting the successes of the company in addressing these issues. Making the workforce feel good about what it is achieving is a very powerful motivator for further innovation.

Suggested outcome—*a structured program for communicating the importance, benefits and opportunities of design for accessibility throughout the workforce.*

Having established the channels for communication and put into place an appropriately nurturing culture for design for accessibility, then the next phase is to build on those foundations to implement changes.

PHASE 3—IMPLEMENTING THE BUSINESS PLAN

The ability of a company to introduce the appropriate changes needed to embrace design for accessibility depends on a number of factors. It requires giving appropriate responsibilities to the people with the right skills. If the current workforce does not possess the right level of expertise, then the senior management needs to have a contingency plan to acquire those skills. This issue is addressed in detail in a later chapter on "Filling the Skills Gap" (Chapter 6).

Stage 3.1—Implement Management Structures for Design for Accessibility

The aim of the first two phases was to learn about design for accessibility in general and then more specifically about how well the company is placed to adopt it. With all that information to hand and a steady increase in awareness and knowledge among the people involved in that learning process, the company needs to begin putting the right infrastructure in place to ramp up adoption of design for accessibility practices.

One key element of the management and, indeed, corporate structure is the communication pathway. In many of these phases, the communication appears to be primarily one-way, being driven down through the corporate structure by the senior management. However, the senior management also needs to ensure that there is a route for information to flow back the other way, from the design teams upwards. One reason for this is a design team may find itself having to balance potentially conflicting requirements where one user group's requirements conflict directly with another group's. A compromise decision may need to be reached and the nature of that decision and thus who makes it may be on of two possible alternatives:

1. it may be a tactical decision, i.e., decided for each individual project, in which case the decision-maker will most likely be the project manager; or,
2. it may be a strategic decision, i.e., impacts the corporate design for accessibility philosophy and program, which would need the decision-maker to be a senior manager.

Tactical decisions are those that affect a single product. Strategic ones are those that affect a significant number of products, maybe even entire product ranges.

Another common decision that may require recourse to the senior management is if the cost increase, associated with developing a design to include a specific set of users, may cross a pre-determined threshold requiring explicit senior management approval. Not only does the senior management need to play an active role in such decisions, it also needs to keep re-examining the design for accessibility approaches and philosophy adopted, if the same issues keep recurring.

Finally, the definitions and code of corporate practice for what is acceptable and what is not acceptable in terms of specifying and meeting accessibility targets should be tightened. as a minimum the definitions and code of practice need to meet all legal requirements from anti-discrimination legislation. Ideally, they should go beyond the level of minimum legal compliance. The accessibility targets specified need to be robust enough so that designers cannot hide behind technical arguments in order to defend their non-compliance.

Anyone who has worked with designers and engineers[1] will know that their first response when asked to do something that they do not want to do is to hide behind obscure technical arguments. Those arguments will be based on the use of jargon to explain and justify why it is not feasible to do what has just been asked of

---

1. Note: I am writing this as a member of both of those groups.

them. As a slightly flippant, but simple rule of thumb, an increase in the frequency of use of jargon in the answer usually corresponds to an increasing desire to dissemble and dodge the issue. Remember that designing for accessibility is a challenge to engineers and designers. Some will want to meet the challenge, others will not. Some will be well equipped to meet the challenge, others will not.

This is not meant to deride the designers and engineers in any way. It is simply a recognition that all "change programs" meet some resistance when being implemented, and the most resistance usually comes from those who are most affected by the changes. In the case of design for accessibility, that is the people who have to change what they do on a daily basis to accommodate a new set of targets and directives, in other words, the designers and engineers on the design team. to overcome the use of technical minutiae as a reason not to make the product more accessible, the corporate targets need to make it clear that the accessibility targets need to be met in all circumstances irrespective of the technologies involved. If the current method of doing something does not support those goals, then alternative technologies need to be examined.

Only if there are no possible alternative technologies, or when the costs involved in those other technologies clearly cannot be justified, should the design be considered for approval—and the decision whether to approve the design should clearly rest with senior management. Remember, these are the criteria that any lawsuit would most likely examine: Could this be done another way? What would the cost burden be? Only if it clearly was not possible for the product to operate another way (using different technology) or if the associated cost was disproportionately excessive, would the company be able to defend a lawsuit on the grounds that the alternative did not constitute *reasonable* accommodation.

At the project level (but chosen and championed by senior management), there are two principal infrastructure options: centralized and distributed teams. The centralized approach involves the formation of a design for accessibility "team." The advantages of this approach are that there is a single point of contact for the rest of the company and that keeping the design for accessibility specialists together should promote and support faster dissemination of new and developing practices within the team. However, the principal disadvantage of this approach is that the apparent separation between the design for accessibility experts and the rest of the company can promote a "them and us" attitude and also, if not managed correctly, lends itself to the "checkbox police" issue discussed in Phase 1. To avoid this potential schism within the company, a proactive agenda promoting collaboration and involvement of the design for accessibility experts must be put into place.

The alternative approach is that of the disseminated model. In this approach, the design for accessibility experts are distributed across the company. The ideal situation would be to have one such person per design team and per marketing team. Under this arrangement, each team would automatically have access to the data and expertise they would need to implement design for accessibility practices successfully. There is also very little chance of the "them and us" difficulty arising. However, the disadvantage is that there would necessarily be a reduced level of interaction between the design for accessibility domain experts than under the centralized model. Thus, slight variations in how design for accessibility was implemented across the design teams may begin to emerge, unless rigorous centralized standards are imposed. If the centralized standards approach is adopted, then a mechanism for policing it needs to be considered.

In practice, a compromise model comprising elements of both the distributed and centralized approaches may prove to be an effective solution. This could be achieved by taking the concept of a single set of domain experts from the centralized approach and then assigning them individually or in small groups to one or more specific projects as per the disseminated model. In this way the advantages of both approaches are available to the company and the potential disadvantages of each are, to a large extent, automatically mitigated.

Suggested outcome—*a clearly defined management structure put in place.*

Stage 3.2—Perform Pilot Studies

When first venturing into the transformation of business processes, it is helpful to perform the business equivalent of experimental pilot studies to help inform and shape the overall transformation process. Thus, in the case of the adoption of design for accessibility practices, one of the first items addressed should be the implementation of case studies to investigate how design for accessibility practices can be incorporated into the particular corporate structure of the company.

It would be a formidable challenge to change all of a company's design and marketing practices overnight or even to change them all simultaneously. At any particular time, there are likely to be numerous products at different stages of development and maturity. Given these different levels, a blanket approach to implementing design for accessibility would most likely not work. Equally, for a company taking its first steps in adopting design for accessibility practices, it is unlikely that the first efforts made in this area will be one hundred percent successful first time. This is likely to be true even if all of the people involved are highly skilled and the very best information is made available within the company. Even industry-wide, well-documented best practices may need a little optimizing to fit within a particular company structure in the best possible way.

Instead of the one-size-fits-all approach, a more measured approach should be taken. One such approach is to identify a range of well-defined projects and specify that those projects should meet particular accessibility targets. These projects can then be monitored to find out which design for accessibility practices worked well and which did not. The practices can then be refined and adapted to meet the needs of the design teams or new practices evaluated if the old ones appear to be sub-optimal for the particular circumstances of the projects.

Suggested outcome—*analyzed results from pilot studies that clearly identify successes and lessons to be learned.*

Stage 3.3—Recognize and Enhance Expertise

$$\text{expertise} = \text{knowledge} + \text{skills} + \text{experience} \qquad (4\text{--}1)$$

In building the level of knowledge required to complete all of the phases and stages this far, the company will have been steadily increasing its collective knowledge and design for accessibility expertise. The purpose of this stage is to identify that knowledge and expertise explicitly and establish channels of communications between the various pockets of expertise and knowledge throughout the company. If we assume that the company wishes to adopt either the centralized or compromise models from Stage 3.2 discussed earlier, then this collection of expertise should be formally recognized as a team in its own right. Doing so would promote intra-team member dialogue and encourage the sharing of domain expertise.

However, the company should not be content with simply sharing its existing expertise across its employees. Once the inherent expertise within the company has been identified, it should be mapped back to all of the previous stages to identify any areas of expertise that are not covered adequately, i.e., skills gaps. Where skills gaps are identified, the company should take prompt action to acquire those additional skills. This is discussed further in the chapter on "Filling the Skills Gap" (Chapter 6).

Suggested outcome—*a corporate map of teams and individuals with design for accessibility expertise and a plan for increasing overall corporate expertise.*

Stage 3.4—Review Progress

Having invested the resources necessary to reach this stage, the principals and assigned executives should review the progress achieved. The purpose for this is twofold. First, it enables the company principals to modify the change program to optimize it to the company's needs. Second, it reinforces the notion that design for accessibility is important to the company. Not only has the senior management researched, implanted and nurtured design for accessibility, but it is also actively monitoring its progress. The message to all employees further down the company hierarchy is obvious—design for accessibility matters to this company.

Suggested outcome—*a review of progress made to date and recommendations for further improvements in the implementation of design for accessibility.*

By the end of this phase, the company will be well on the road to full adoption of design for accessibility practices in product development. While the actual mechanics of implementation may need the occasional tweak to maintain the best possible fit, the senior management can now turn its attention to developing a full-blown corporate program to make sure that design for accessibility becomes central to everything that the company produces. This will most likely involve re-visiting many of the stages addressed so far and, instead of looking at individual projects on which to try out design for accessibility, start looking at rolling it out across entire product ranges and families, until all of the company's output is designed to maximize its accessibility.

PHASE 4—SELLING ACCESSIBILITY

The final stage of implementing design for accessibility is to begin to use it as a differentiator in the marketplace. It is during this phase that the company should begin to see competitive payback for all of its efforts.

Stage 4.1—Identify and Leverage Competitive Advantages

Implementing design for accessibility involves a wide range of variables that require balancing. Different companies will find different solutions as to how to achieve the correct balance for them. Some of those solutions will achieve more success in the marketplace than others.

As a company becomes more mature in its design for accessibility approaches, it can begin to fine-tune and finesse its market offerings. Elements that worked well and proved successful for one product can be adapted and applied to other products. Less successful features can be replaced or revised. Brand new product ranges and brands may evolve out of the original, creative work generated through the adoption of design for accessibility practices. The company may also be well positioned to move into completely new product areas where design for accessibility is not being applied. An excellent example of this is how OXO grew its GoodGrips brand from what was basically a well-designed handle on a very limited number of products (potato peelers, etc.) to a range of over 750 products (CDF, 2001).

Suggested outcome—*a structured plan to transfer the successes in design for accessibility throughout the company products range and brands.*

Stage 4.2—Identify Opportunities for Improved Corporate Image

Design for accessibility is a very user-centric design discipline. A product that has been designed using design for accessibility principles and practices is extremely unlikely to exhibit significant usability, accessibility or utility problems. In other words, a product design using design for accessibility practices is highly likely to be a well-designed product.

As such, this is something that a company should seriously consider promoting in the public domain. However, looking back at the earlier discussion on branding issues, not all companies will want to advertise that their products were designed for someone who had, for example, difficulty unscrewing a jar lid because of arthritis in their fingers. Instead, concepts such as "easy-off" or "designed for you comfort" should be considered. Remember, effective design typically stands out in the marketplace. It does not need to advertise where the good design came from or how it came about, just that it *is*.

Additionally, not all of the market-place benefits of adopting design for accessibility are financial. Companies also stand to enhance their reputation for their focus on good design. At the time of writing, Apple is still dominating the mobile personal music market with its various incarnations of its iPod concept. That dominance has arisen through a happy coincidence of the right product at the right time, accompanied by the right marketing approach. The iPod has beaten off its much cheaper competitors through an astute combination of form and function. It looks "cool" and customers love its control interface.

Since the success of the original iPod, Apple has been seeking alternative avenues for taking advantage of its current prominence in the marketplace. Not only is it developing new products, such as mobile telephones that also play and download music, it is also partnering with other companies with a reputation for excellence in design, such as Bose and JBL, to make and market larger sound systems based on Apple's core iPod technology. With the "next big thing" in computing likely to be centralized home entertainment systems driven through a single central computer, Apple seems to be maneuvering itself for a launch on that particular market. This strategy could potentially ensure Apple's survival for years, assuming that it is not undone by trying to milk the market too fiercely and sacrificing quality control in the pursuit of larger profit margins.

Underlying all of Apple's strategy is the customer goodwill that has been generated through the iPod concept and, if this goodwill is not squandered, could stand the company in good stead for quite some time. Of course, this goodwill did not just appear overnight. Apple has been carefully managing its marketing campaign to achieve this result. It is also trying to transfer some of the luster, also known as the "halo effect," (i.e., reflected glory) from the iPod to other parts of its consumer range, notably its laptops and desktop computers. Apple appears to be having some success in achieving this goal (Wong, 2005).

This example demonstrates how developing a product range that meets the customers' needs and aspirations can become the proverbial goose that laid the golden egg. Apple's attempt to promote its other products on the back of the iPod's market luster is the type of activity that a company could realistically engage in if it develops a really successful product that has been designed for improved accessibility. Luster is a valuable commodity. It is intrinsically connected with branding and brand recognition. Many of the user-centered design practices associated with design for accessibility lend themselves well to creating a product that will generate luster in the marketplace.

Suggested outcome—*a structured marketing plan for communicating the new corporate and brand image.*

PHASE 5—REVIEWING AND REFINING BUSINESS PLAN

In reaching this final phase, the company will have succeeded in testing the waters with regard to adopting design for accessibility and acquiring a broad range of in-depth knowledge. The final phase is to examine the progress made up to this point and to review and refine the business plan to maximize the company's effectiveness in designing for accessibility.

Stage 5.1—Recognize and Reward Successes

The achievements of the company in embracing design for accessibility should also be celebrated within the workforce. Looking back at Phase 2, where we examined how important it was to ensure that everyone within the company was fully on board with the idea of adopting design for accessibility practices, now is the time to communicate those successes throughout the company. Senior management should take this opportunity to highlight products that achieved particular success at meeting product accessibility targets, gained significant market-share or received notable public acclaim or even prizes. It would also be worthwhile highlighting instances where the design for accessibility approach achieved beyond what would have been expected of the company's previous design approach.

Showing the workforce that its dedication and effort have been recognized and valued will go a long way towards inspiring further developments and advances in the company's design for accessibility practices. There are numerous ways to acknowledge a particular team's achievements: internal prizes and rewards, financial and otherwise; honorable mention on the company's Web site; celebratory dinner at a good restaurant. Senior management has the opportunity to be as inventive as its design teams in rewarding their success.

Suggested outcome—*a reward program that recognizes and encourages innovation.*

Stage 5.2—Review and Refine Design for Accessibility Approach

Having reached this stage, the senior management will have garnered a wealth of information about how successful, or otherwise, the adoption of design for accessibility practices has been. Throughout the phases and stages that preceded this one, senior management will have been actively engaged in review cycles, but those will have been primarily tactical in nature, i.e., of the sort that asks the question "How are we going to tackle this phase/stage?"

This is now the time to perform a more strategic, global review that asks questions along the lines of "Did we meet our initial objectives?" and "How can we do better?" To answer these questions, the senior management needs to organize and co-ordinate a proactive review cycle that looks at all aspects of the product development cycle. Critical comments should be welcomed and not discouraged. If this only serves as a collective pat-on-the-back for a job well done, then that will only promote corporate complacency. The company needs to be actively, and even aggressively, seeking opportunities to improve its design for accessibility approach.

Suggested outcome—*a completely realized infrastructure for managing and implementing design for accessibility.*

At the start of this chapter, design for accessibility was compared to design for safety, because of the similar need to place design for accessibility in the very center of the design process. The analogy continues through to this last stage. There is no such thing as a perfectly safe, product. Similarly, there is no such thing as a perfectly accessible product. Just as there is always a way to make something that little bit safer, there is also always a way to make something that little bit easier to access and use.

KEY POINTS

- Top management plays a pivotal role in implementing design for accessibility practices and a continuing role in maintaining design for accessibility practices.
- Top management initiates and drives the initial adoption of design for accessibility and retains ultimate responsibility for the success of design for accessibility.
- Top management shapes the company's design for accessibility philosophy and is responsible for communicating this throughout the company and ensuring that corporate targets for product accessibility are met.

5

The Role of Project Management

Below a company's top management, but above the designers involved in the day-to-day design process are the middle level executives who typically commission and administer the product development projects. Middle management requires a delicate touch to maintain the correct balance between the stated corporate objectives driven by senior management and the practical limitations faced by designers in trying to develop products to meet potentially conflicting design requirements. The aim of this chapter is to help provide project managers with the material to establish more inclusive design briefs and also provides mechanisms for reviewing the progress of the product accessibility and keeping the design teams focused and motivated.

As with the preceding chapter on the role of senior management, this chapter is not intended to provide a complete description of all of the issues surrounding the adoption of design for accessibility. It is a starting point for inspiring further thought. Not all suggestions will apply to an individual company's circumstances and in many cases some degree of adaptation and modification may be required to get the best fit.

The tools and techniques available to assist middle management in implementing design for accessibility are best illustrated within an example design process. There are many such processes available within design literature. For our purposes here, we will assume that a hypothetical company wishing to adopt design for accessibility practices has a fairly generic design process that features all of the common design process elements, as outlined below.

A DESIGN FOR ACCESSIBILITY PROCESS

The design process we will use here is split into four phases, each with up to four component stages, as shown in Table 5–1. An important feature of this design process is the enforced use of strict gateways at the end of each stage and each phase.

TABLE 5–1
A Design for Accessibility Process

PHASE 1—Define Project
Stage 1.1:   Initial research
Stage 1.2:   Develop design brief

PHASE 2—Design, Detail and Implement Solution
Stage 2.1:   Generation of solution concepts
Stage 2.2:   Selection and refinement of most effective solution
Stage 2.3:   Detail design of solution
Stage 2.4:   Ready solution for production

PHASE 3—Go to Market
Stage 3.1:   Launch of product in marketplace
Stage 3.2:   Evolution of product
Stage 3.3:   Extension of product range

PHASE 4—Project Closure
Stage 4.1   Decommissioning of product
Stage 4.2:   Final review and lessons learned

The gateways are effectively review processes with a number of potential outcomes including:

- approval to proceed to the next stage;
- referring the design back to an earlier stage for re-working; or,
- even stopping the project completely.

The decision whether to proceed or not is based on a review of the work in that stage, and earlier stages where appropriate, confirmation that the accessibility properties have not been compromised and that the business context is still intact.

PHASE 1—DEFINE PROJECT

The purpose of the first phase of the design process is to understand better what the project is trying to accomplish and whether it is technologically and financially feasible. Assuming that it is feasible, a project plan needs to be developed that will form the foundation of the following phases.

Stage 1.1—Initial Research

Once a potential project has been brought to the attention of a project manager, the first priority is to establish exactly what the opportunity is and what it represents. This process would typically involve establishing what the customer needs and aspirations are that are being targeted and whether any solutions currently exist to meet those requirements (and how well they are being met). The aim is to develop an understanding of the current market situation and an initial outline of how the opportunity may be tackled.

The adoption of design for accessibility practices into the design process requires the addition of a few extra pieces of data to collect and consider. For example, the project manager needs to identify any specific legislative or standards requirements that have to be met for that particular marketplace. It is also necessary to see how the opportunity relates to the corporate inclusive design philosophy. With design for accessibility still being a comparatively new field in design, one area of opportunity that has to be investigated is which market segments are simply not being served at all by the current solutions. The first company to reach those segments should stand to gain a large share of that market.

It is also necessary to examine how heterogeneous the target customer base is and what user groups are present within it. Having identified the user groups, these should then be looked at in the broadest possible terms to identify the full range of user capabilities that may be encountered. For example, even if the user group is defined as "active 20-somethings" this may include women who are pregnant, people with broken bones or other sports injuries (some 20-something activities can be downright dangerous) and so on. The more imaginative the interpretation of the user group at this stage, the more accessible and inclusive the final design is likely to be. Also, remember that sometimes products designed for one user group find a niche serving another group. Most modern sneakers are designed for younger adults playing sports (or for just looking cool on the urban streets). However, those with air-cushioned soles are also popular among older adults with arthritis in their knees, because it reduces the impact and pain when they walk. Similarly the use of Velcro-type fastenings instead of traditional laces on the sneakers is also easier on those arthritic fingers.

Suggested outcome—*a description of the opportunity.*

Stage 1.2—Develop Design Brief

The feasibility of developing a product to meet the opportunity relies on:

1. appropriate technology being available to meet the customer needs in a cost-effective manner;
2. the perceived market demand being real; and,
3. the company being in a suitable position to respond to the need, i.e., having the resources, skills and expertise available at the right time.

Once it has been confirmed that there are realistic potential solutions to each of the above issues, those solutions can then be included along with the description of the opportunity to create a project proposal.

The addition of design for accessibility to the design process affects each of these issues. First, let us look at the availability of appropriate technology. For most product development cycles, the existence of a single technological solution for a problem may be sufficient. When considering design for accessibility and the concomitant increase in diversity of user capabilities, though, there most likely needs to be multiple methods (or modes) for achieving the same goal. For example, a technological solution that is dependent solely on speech will almost certainly not work for someone who is deaf. An alternative mode of operation is required, such as allowing the use of text. Ideally, the technological solutions identified will be:

- robust, i.e., capable of coping with a wide range of user capabilities;
- flexible, i.e., able to support multiple ways of achieving the goal); and,
- scalable, i.e., able to be combined with other technologies to achieve the goal.

Next, confirming the perceived market demand would ordinarily involve market surveys, user focus groups and maybe even the testing of early prototype solutions. The addition of design for accessibility objectives at this stage requires that the participants recruited for those market research activities truly represent the interests, wants and needs of the full range of potential users. The choosing of participants for this stage echoes many of the issues regarding the selection of users for user trials discussed later in Chapters 9 and 10. Note that the ideal participants at this stage would include users. However, if suitable users cannot be located or recruited for logistical reasons, then it is better to include user representatives (e.g., carers and occupational therapists, etc.) than to use no one at all. Having said that, users are the very best resource for finding out about what they want and every effort should be made to find some real users to participate in the market research activities.

Lastly, regarding the company's ability to respond to the need, the availability of design for accessibility expertise that can be offered to the design team needs to be considered. Particularly in the early stages of adopting design for accessibility, most companies will need to ensure that they do not over-extend their design for accessibility team. The design for accessibility component of the design process cannot be shrink-wrapped and inserted at will. It needs to grow with the design solution and, consequently, the design for accessibility experts need to be involved throughout the full length of the design process. A further consequence of this is that the allocation of those experts' time needs to be carefully considered and planned for the company's maximum advantage.

The final decision on whether the project is feasible will most likely draw on advice from marketers, designers, the relevant administrative officers and the design for accessibility team. The decision itself will be made either by the project manager or, if a large-scale strategic project, possibly by the senior management.

Suggested outcome—*a design brief that states the general objectives and requirements of the project.*

PHASE 2—DESIGN, DETAIL, AND IMPLEMENT SOLUTION

If the project is considered feasible and approved, then the next phase is to design, develop, and realize a final complete product. As with all of the senior management phases, it is imperative that detailed records are kept of all design decisions and this is especially important when considering which design alternatives are kept and which are discarded. Not only are such records invaluable should a claim ever by lodged against the company, they are also excellent information resources for future designs. As the old adage goes, there is no point re-inventing the wheel. Records of why particular design options were discarded may be transferable and relevant to the design of other products, thus resulting in a streamlining of their respective development processes.

In addition, since technology typically develops rapidly, design options that were discarded in the past because they were deemed technologically infeasible may become more realistic options at some point in the future. Keeping those options on file would help future design teams, saving them the effort of having to develop those options again from scratch, i.e., not having to reinvent the wheel.

Stage 2.1—Generation of Solution Concepts

The first step in the development of potential product solutions is to re-examine and finalize the project specification. This process involves deciding the preferred utility (functionality) of the product and the design priorities (the specific targets to be met and their comparative levels of importance).

Once the specification has been completed, the design team can begin the process of generating potential solutions and evaluating them against the project specification. The final step in this stage is to decide which solution is the most promising to take forward to the next stage. At first sight, the principal impact of design for accessibility on this phase would appear to be in assisting the weeding out of design options that exclude particular user groups. This weeding out could be achieved through simple user trials incorporating users that reflect the full range of capabilities that are found in the target customer populations.

For example, imagine that the product being developed is a mobile telephone and that it is targeted at adults. Under the principles of design for accessibility, it would be perfectly reasonable to test the product with an older lady with arthritis and slightly failing vision. Doing so would quickly reveal that the buttons should not be too small or placed too closely together for that user. At the same time, the text labeling on the buttons and on the screen should not be any smaller than a particular font size. Any design options breaching those limits would not be accessible to that older lady and millions of others like her.

Thus the adoption of design for accessibility practices at this early stage of the design would help the company identify and avoid design options that would later need rectifying or possibly even lead to the abandonment of a product concept after a significant amount of developmental expenditure. This risk is especially great if the problems remain undetected until later in the design process. However, this is not the only role that design for accessibility practices can play at this stage. They can also help with the generation of potential solutions. For example, user observation sessions in which designers watch users with particular functional impairments perform a task can be an inspiring activity in its own right. Designers are very often "visual" people and seeing something first-hand is a great way of getting their creative juices flowing.

Additionally, there is the hidden benefit of coping strategies. Coping strategies are the novel techniques that users develop to overcome a particular difficulty using a product, or accomplishing a task. These are often very creative solutions and are a great resource for designers wanting to develop radically new alternatives. A good example of how users think laterally is the popularity of mobile telephones among people who are deaf. Initially, most people's response to learning this is one of bemusement—why would someone who cannot hear want a mobile telephone? The answer to that is not for the telephone itself. Instead, it is for the ability to send text messages and e-mails. Equally, many mobile telephones offer calendar software. Combine that with the ability to vibrate (instead of just ring) and a deaf person has a mobile personal organizer with a built-in alarm that they can feel. Suddenly the popularity of mobile phones among the deaf community is no longer a surprise; it makes perfect sense.

Returning to the main objective of this phase, to select the most promising design solution, it is possible that the range of solutions identified meet the needs of some sections of the population, but not others. It may not be technologically possible to meet the needs of all of the potential users. Under these circumstances, the decision as to which solution will become the preferred solution to be developed further will need to be taken. The design for accessibility criteria on which that decision is based should include:

1. the company's stated design for accessibility philosophy and targets;
2. the respective levels of accessibility of each of the potential solutions;
3. how those levels of accessibility map to the demographics of the target customer populations; and,
4. whether there are any explicit legal accessibility requirements that need to be met by the product.

These criteria should be considered in addition to the usual ones that the company evaluates as this stage of the design process, such as overall functionality, etc.

Suggested outcome—*a range of potential alternative solutions that meet the design brief.*

Stage 2.2—Selection and Refinement of Most Effective Solution

By this stage, the chosen solution will still be at a comparatively abstract level of design and needs to be developed into a more tangible and concrete product. All manner of constraints will need to be considered. Technical limitations imposed by what the current or anticipated technology can achieve may limit, for example, how much water a coffee-machine can heat on a 120V supply. Financial limitations may dictate that the chrome metal finish on the coffeepot may have to be replaced by a plastic alternative. Safety limitations may mean that the lid of the pot cannot be removed by a child when it is full of scalding hot water and so on.

From a design for accessibility perspective, having weeded out the design solutions that do not meet the company's design for accessibility objectives, the aim now is to ensure that the design for accessibility elements of the product are not forgotten in the mass of other requirements and constraints as the solution develops. An effective technique for achieving this is to ensure that user trials with the preferred solution continue throughout its development. If this is not economically or logistically possible (e.g., the users are too remote to work with on a frequent enough basis), then a minimum requirement of a final *successful* user trial should be established before the design is allowed to proceed to the next stage.

In the meantime, prior to the final user trial, design for accessibility experts should continue to be consulted throughout this stage by the product design team. Those experts could be members of the company's own design for accessibility team, or external consultants and should ideally be full-time members of the product design team.

Suggested outcome—*a solution that meets the design brief most effectively.*

Stage 2.3—Detail Design of Solution

By the time the product reaches this stage, it is well on the way to becoming a commercial reality. All of the functionality should be present and all that remains is to fine-tune the design so that it is ready to pass to the people who will be responsible for manufacturing it. This stage involves checking that the product meets all of its design specification in terms of utility (functionality), usability and accessibility. As with the preceding stage, the design for accessibility aim at this point is to ensure that the additional details added to the design do not compromise the underlying accessibility integrity of the product. User involvement, most likely in the form of participation in user trials, is the most obvious method of achieving this. Again, the use of design for accessibility experts can be used to substitute for the presence of users for the majority of this stage, but the final design should be verified by a mandatory final user trial.

A business decision may also need to be made at this stage. By now, the product will be a very tangible high-fidelity prototype. All of the functional elements will be in place and most of the aesthetic ones as well. The project manager, and possibly someone from the senior management, may need to review how well the new product fits into or complements the existing product range and branding. If it represents only a small deviation from the rest of a well-established range, then it may be decided to keep the product as part of that range and brand. However, if it becomes clear that the product has developed into something quite new and different, then the company needs to decide whether to keep the new product within the current branding and re-align the brand, or develop an entirely new range and brand. This is an important decision, especially for the first few products that a company develops using design for accessibility, because the resultant outcome will affect how the company positions itself and its brands within the marketplace.

Suggested outcome—*a detailed design of the chosen solution.*

Stage 2.4—Ready Solution for Production

At this stage, the overall product functionality and style will have been defined and put into place. The only changes will be ones dictated by the manufacturing process and the final stages of the application of branding. As for Stages 4 and 5, those changes have to be checked to ensure that they do not affect the overall accessibility of the final product. The mechanisms implemented for the two preceding stages can be applied here. This is also the stage when the full product packaging will be developed. The packaging itself needs to be as accessible as the product, in other words anyone who can use the product needs to be able to get it out of the box that it is sold in. Labeling needs to be clear and readable. In fact, many companies may feel that the entire issue of packaging design is worthy of its own design for accessibility study.

User guides and product manuals are another important issue. They need to be both readable and understandable by all potential users, which means the use of larger fonts than typically seen, especially for safety warnings. To reduce costs, many companies are making their product manuals available on-line only. This approach is a mixed blessing for some users. On the plus side, a person who is blind may be able to use a screen-reader to access the contents of the manual. However, for that to happen, the company Web site needs to be accessible to all purchasers and the manual stored in a format that the screen-reader can interpret correctly. It also requires that the users have a readily available Internet connection, plus also maybe a printer if the manual is quite lengthy. Consider if the user cannot get on-line and needs to telephone the customer support number, only to discover that the telephone number is in the on-line manual. This is, unfortunately, a surprisingly common occurrence.

Suggested outcome—*a production-ready solution.*

PHASE 3—GO TO MARKET

The company now has a product that, in an ideal world, meets all of its design for accessibility targets. In practice, it will more likely have a product that meets as many of its targets as current technology and cost constraints allow. The next stage is to launch the product in the marketplace.

Stage 3.1—Launch of Product in Marketplace

Launching any product requires careful planning and coordination of multiple teams within the company:

- the design team has to develop the right product for the right time;
- the marketing team has to convince people to buy the product as well as manage and enhance the company's reputation through the luster surrounding the new product;

- the sales team has to generate orders and revenue from the product; and,
- the customer support team has to be able to help purchasers who encounter problems with the product.

Marketing the product presents an interesting set of choices to the company. Most companies will naturally want to exploit the accessibility of the products that they have developed after devoting perhaps considerable time and effort in developing those attributes. However, there is no one-size-fits-all marketing approach and a blanket promotion of how accessible (and thus inclusive) a product is may not always be the best solution.

The overall aim of design for accessibility aspects of the marketing campaign will usually be to emphasize how natural and straightforward the product is to use. Note that the word "easy" was not used here. "Easy" is a good word when describing a product that has a reputation for being complicated. Under those circumstances, most people would actively look for something that made it clear that it was not difficult to use. However, there are times when "easy" comes across as patronizing and almost demeaning. The joy in completing a word puzzle or jigsaw puzzle that is fiendishly difficult is so much more than one labeled "easy." Similarly "comfortable" can be a useful word when describing a toilet that has been designed for improved accessibility. However, it is regarded with extreme suspicion when describing the latest clothing fashions, for example, where "exclusivity" is sometimes considered a desirable attribute.

The sales and marketing teams have a big role to play in ensuring that the right tone is taken to maximize the market acceptance of the product. It may even be worth developing multiple marketing campaigns targeted at different sectors of the population, each with their own slightly different emphasis on specific product features. For example, while a car with doors that open wide may have been designed for someone with arthritis to make it easier to get into, the same feature can be marketed to new parents as great for lifting infant seats in and out. As with most things in life, it is all a matter of perspective.

Suggested outcome—*a carefully planned and executed launch strategy.*

Stage 3.2—Evolution of Product

Very few products are launched into the marketplace and then left to fend for themselves. Those that are effectively left alone are typically low value products whose designs have not changed for many years, clothespins for example (clothes pegs in the UK). Having invested the time and effort to develop a product, companies will seek to maximize their return on that product. Common strategies for sustaining the product in the marketplace include the periodic addition of new features and cosmetic facelifts.

When it comes to design for accessibility, the opportunity exists to examine whether the product is presenting unexpected difficulties to any of the target user groups. This does happen—it is almost impossible to design for every possible contingency, especially when dealing with such a diverse user population. However, that does not excuse not making the effort to do so! Once such difficulties are identified, the on-going re-design specification for the product should be amended to include fixing the causes of those difficulties. This is precisely in line with the concept of countering design exclusion—find the difficulties and then remove their causes.

Suggested outcome—*a series of product updates and augmentations based on customer and market feedback.*

Stage 3.3—Extension of Product Range

Maximizing value from a product will most commonly be interpreted as maximizing monetary income from it and the most common interpretation of this is sales revenue. However, this is only one aspect, albeit an important one, of the concept of value of a product. Products are rarely developed and launched in isolation. They are often part of an existing range of products. If they are not, and later prove successful, a range of products will almost certainly be developed around them. Apple's iPod is an example of this. One of the important benefits of the use of ranges is that the company can offer variants on a product that are tailored to meet the specific needs of different market segments. This capability to fine-tune the product offering without having to repeat the whole product development process is financially attractive. The all-important proviso here is that the fine-tuning should *not* be in the form of bolt-on additions, but rather, instead, integral functionality added to the underlying concept.

Additionally, there are positive opportunities to expand the sales of the product to other user groups not originally envisaged as potential customers in the original product specification. The marketing of mobile telephones to the deaf community, as discussed earlier in Stage 3, is an example of a type of product reaching out beyond its original target population. The marketing team should be actively seeking out such additional markets. Finally, legislation rarely remains static for long, especially when considering anti-discrimination legislation. The company needs to remain alert to any changes in mandated accessibility targets. In particular, it must be noted that much legislation is based on the concept of reasonable accommodation. This is important because it is likely that potential design solutions were discarded during the development of the product because it was not technologically possible to implement them.

With technology changing all of the time, what was once technically infeasible may at some stage, become possible to implement. So, for example, it is possible that a technological advancement could make a task or process accessible that had, up to that point, been inaccessible for users with a particular set of functional impairments. As soon as that happens, all companies claiming that they could not accommodate the needs of those particular users, because it was not technically possible to do so, will be obligated to amend their products (or introduce new ones if circumstances permit) to meet the needs of those users. Otherwise those companies run the risk of a possible legal action against them. Thus, the company needs to remain actively on the lookout for such technological advancements. Ignorance of them is unlikely to be a satisfactory defense when on the wrong end of a lawsuit.

Suggested outcome—*a series of new or modified products to complement and build on the success of the original product.*

PHASE 4—PROJECT CLOSURE

Implementing design for accessibility within a company represents a long-term commitment to realigning the design process such that it is centered on the needs and aspirations of the customer/user. As such, the impact that design for accessibility has on a company's products needs to be reviewed and analyzed all the way through to the end of each project, which includes the final decommissioning of the product and withdrawal from the marketplace. Once that stage is complete, a full review of the project across its entire lifetime is possible.

Stage 4.1—Decommissioning of Product

The final stage of the life of the product is its withdrawal from the marketplace. For many products this can be a sharp and brutal removal, especially in the consumer electronics arena where products have a shelf life measured in weeks and months, rather than years. However, more sensitivity may be required when considering products used by segments of the population such as people with disabilities or older adults. At the risk of over-generalization, many of those users may not be as accepting of frequent changes of models and products. They often aim to purchase a product and keep it for a long time, sometimes becoming very attached to it. The arbitrary removal of that product from the market and, more importantly, the removal of support for the product may not be well received.

Consequently, the withdrawal of the product may need to be managed and orchestrated as carefully as the launch. This is not to say that a company should never withdraw a product; that clearly does not make sense. Instead, the company should think about the customer over a longer time frame than simply one product's lifespan. Product ranges do not just exist at this point in time. They extend backwards into the past and forwards into the future. The company's aim should be to evolve the product and the customers' expectations at the same rate, taking the customers along for the journey, rather than dumping them unceremoniously at the roadside.

One way of achieving this is to retain many of the key features and attributes that the users have grown accustomed to over time and slowly add in new functions, but at a rate that never overwhelms the user. In this model, a product is seldom really withdrawn, but continually evolved into newer products.

Suggested outcome—*a carefully planned and executed market withdrawal strategy.*

Stage 4.2—Final Review and Lessons Learned

Valuable lessons can be learned from a final review of the project and, if handled properly, those lessons can inform and improve the company's overall design for accessibility approach. Typically a final review of the project focuses on documenting and archiving all of the stages of the project. It also represents an opportunity to examine the successes and shortcomings of the project in light of the overall success of the product in the marketplace. That success is usually judged primarily in terms of financial return on investment, but can also include favorable reviews and buzz generated by the product and whether any of that luster is, or can be, transferred to other items in the product range. Market research activities for obtaining feedback from customers are valuable methods of obtaining information about the overall product "experience."

Many of the lifetime review objectives apply to the design for accessibility aspects of the product. It may be difficult to tease out what proportion of the revenue generated was directly because of the product's increased accessibility. Instead, it may be better to focus on identifying whether design for accessibility provided any added value to the overall product experience. Customer feedback that includes comments with phrases like "easy to use" or "so simple" is a good indication that the improved ease of access was a factor in the product's success.

Armed with a diverse set of data where the product succeeded or came up a little bit short, the company should review where improvements could be made in its approach to design for accessibility. Specific lessons such as "this feature simply did not work" or "users found the instructions too complicated" should be entered into whatever data repository the company is using to store its acquired knowledge about design for accessibility. Where specific instances of failures are identified, they should be examined to determine whether they were the result of the following factors:

- *Constraints imposed by the technology available at the time of manufacture—and whether those constraints still apply.* If so, then a new way of offering the functionality is required to avoid those same problems with the next generation of products.
- *Additional factors not considered by the design team.* It is simply not possible to consider and design for every potential type of loss of capability within the customer base, or every single customer need. Some will have been overlooked and if those proved to be important oversights, then they should be flagged for the next iteration of product development.
- *Flaws in the design for accessibility approach adopted by the company.* Design for accessibility can be quite complex to implement. The company is unlikely to get it right immediately and mistakes will almost certainly be made. The important point is to recognize when mistakes have occurred, find out why and then fix the cause.

Suggested outcome—*a comprehensive final review of all aspects of the full life of the product with clearly identified successes and lessons to be learned.*

CHAPTER FIVE

KEY POINTS

- Project managers must embrace the concept of design for accessibility if it is to be implemented successfully within the company.
- Project managers are responsible for ensuring that the design team meets the design for accessibility targets set by senior management.
- Document everything. Written records of why particular decisions were taken are the basis of an invaluable knowledge resource.

6

Filling the Skills Gap

The preceding three chapters discuss the need for companies to implement design for accessibility. Throughout those chapters it should also have been clear that successful design for accessibility requires specific skills and expertise to achieve.

If we accept the assertion that "designers design for themselves," unless directed to do otherwise (Cooper, 1999), it is quite clear that a company attempting to implement design for accessibility for the first time cannot simply ask its existing designers to suddenly start designing more accessible products. This is a complex area to work in, with practitioners often requiring at least some knowledge of usability engineering practices, design, cognitive and behavioral theory, physiology and the like. The practitioners also need to have the ability to understand, empathize with, and communicate the needs of the users to other members of the design team. With no widespread educational programs in place to develop potential employees with the required skills, companies need to plan their own strategies for acquiring the information that they need both in the short-term and the long-term. This chapter presents some of the most common solution options that companies adopt to address this challenge.

OPTION 1—USE TOOLS

There are numerous tools available to assist design teams trying to develop accessible products and most of them are computer-based.

Compliance Testing Tools

For instance, computers are excellent tools for storing and processing data, as well as collected wisdom. As such, companies involved in regular, repeatable design activities are likely to find some form of computer-based support available. A good example of this is found in the Web site design community. Tools such as Watchfire's WebXACT (formerly CAST's Bobby tool) are available to perform checks on a

Web site's compliance with the W3C's Web Accessibility Initiative guidelines (Watchfire, 2005). These tools have been evolving over a number of years and are continually becoming ever more sophisticated.

Having said that, automated tools still have limited "intelligence." Consequently, their application is often limited to only the situations that the software developers foresaw when developing the tools. Even in these circumstances, they may still fall short of meeting their objectives. For example, most tools for authoring web pages or checking the accessibility of Web sites have some capacity for checking the existence of ALT TEXT for images (the text that is read aloud by a screen reader to a user with severe vision impairment who cannot see the original image). However, those tools are unable to check the relevance of the ALT TEXT provided. With some images, it may be appropriate for the ALT TEXT to be left blank (where the images contain no relevant information—such as blank spacer .gif files). In other cases, a quite detailed description may be required. Of course, if the description exceeds a certain length, then it may be necessary to point to a LONGDESC attribute.

Visualization Tools

Computer-based tools do not just focus on compliance testing. There are several free tools available from the Web for simulating the different types of color blindness, such as Vischeck (2005). These can be used to perform quick checks to make sure that the colors used on an interface can be discerned clearly by users with varying forms of color blindness. More sophisticated tools can suggest alternative color schemes that increase contrast, and thus readability, for many users. However, human judgment is required to ensure that the suggested schemes retain or achieve the desired level of aesthetic pleasure.

The development of visualization tools is continuing apace. The latest generation of web accessibility assessment tools, such as IBM's aDesigner (IBM, 2005), not only offer compliance testing against a range of standards such as WAI (W3C, 2005) and Section 508 (WIA, 1998), but also offers the ability to simulate the effects of a range of visual impairments. This includes the different types of color-blindness, reduced visual acuity and also a visual representation of the time taken to reach a particular part of the screen display when using a screen reader. The visual representation is based on the fact that a screen reader will always read a page in a particular order, effectively starting at the top of the page source code and working its way down through the page. The representation itself works by coloring anything reached quickly, say in less than 5 seconds, as white. Anything that takes what is considered to be an unacceptably long time (the default is 90 seconds) is colored black and everything else between those two limits, is represented in increasingly darkening shades of gray.

Such an approach affords a designer a very fast overview of how long it would take to reach all the important elements on a page. It also encourages a restructuring of the page so that the important content (such as a news article on a news site) is reached very early. These tools are the beginning of a new generation where mere access is not sufficient. Usability—here in the form of time to reach the important content—is also being tested. It is highly likely that HCI tools will continue to develop to look increasingly for both of these attributes: accessibility and usability.

Modeling Tools

Other types of computer-based tools include anthropometric (or biometric) modeling systems. These systems typically consist of three-dimensional articulated models of human skeletal features (e.g., the bones in someone's fingers) combined with data on joint ranges of movement (e.g., the angle through which a wrist can rotate) and sizes of limbs (e.g., how big a thumb is). Some systems exist, and others are under development, that model how someone's arms and hands move, for example, when attempting simple reaching tasks. There is a clear role for such software when checking and verifying, for example, whether there is sufficient clearance around a handle for someone to pick up a kettle, or whether someone can reach an object on a shelf. Such software is often faster and easier to use than plowing through the standard anthropometric reference texts (e.g., Peebles and Norris, 1998).

However, these anthropometric modeling systems are based on assumptions. They can only test an interaction based on how it has been modeled. Real users are often unpredictable in how they accomplish a particular task. a biometric model of upper torso movement will usually presume that the user does not move from one particular location, whereas, in practice, the user may simply take a side step to move nearer to an object.

For anthropometric modeling to be universally applicable, it is necessary to model the full range of human motion *combined with* a sophisticated mental model of human problem-solving techniques. Both of these requirements represent significant research and development challenges. While the locomotion of the basic human skeleton and muscles are quite well understood, they combine to make a system with almost unlimited degrees of freedom of movement. This is especially true when the abilities to move within a specified environment are combined with those of making use of supporting technologies (e.g., standing on an up-turned bucket to reach the object on the shelf discussed above). Additionally, humans are remarkably adept at finding unusual ways of accomplishing particular tasks. Capturing and modeling that variety of approaches is an extremely complex task. For example, it is highly unlikely that an anthropometric modeling program would even be aware of the option of standing on a bucket to reach something that was otherwise too high, yet this is an eminently pragmatic solution to anyone capable of standing on a bucket.

Tools Summary

In summary, computer-based tools are capable of highlighting where there is an actual or potential deviation from an accepted and anticipated norm, but they are usually unable to quantify how significant that deviation is, or to suggest possible remedies. This is especially true when considering how a potential, real user may respond to a particular set of circumstances. However, these are, and will remain for the foreseeable future, complements to detailed expert analysis. They are still too limited to replace the expert. Expert input is typically required to set-up the tools in the first place and for interpretation of the results once the tools have finished their analysis.

OPTION 2—BRING IN EXPERT CONSULTANTS

A common solution for companies looking to benefit from high levels of specific expertise without having to recruit potentially expensive employees is to hire them on a project-by-project basis. This is why consultants exist.

Benefits of Employing Consultants

Consultants generally offer effective, highly focused expertise and an ability to deliver under pressure and often-restrictive time constraints.[1] Another benefit of employing external consultants is that their opinions are often valued more than those of employees within the company, irrespective of whether their opinions are more. Sometimes a perfectly sensible proposal put forward by one group of employees may be dismissed by other employees simply because of past internal rivalries. This is an illustration of the old saying "familiarity breeds contempt." However, should an external expert subsequently propose a solution that is virtually identical to the one that was previously dismissed, it will often be considered seriously as the consultant is seen to have no ulterior motive.

Drawbacks of Employing Consultants

However, there are potential drawbacks to employing consultants. For instance, there is no recognized national or international professional body that audits the qualifications and experience of those who work in this field as a basis for accrediting them. Therefore, clients need to check the credentials of prospective consultants carefully. The age-old adage *caveat emptor* (buyer beware) still applies. Simple, commonsense checks are useful, such as asking for a list of other clients with whom the consultant has worked and taking up references.

Another potential drawback in using consultants is that client companies are unlikely to gain any adequate competency in design for accessibility to apply independently in subsequent projects. This shortcoming can be avoided if the clients incorporate an element of education when commissioning contracts.[2] Without

---

1. I, and indeed many others in this field, spend a significant amount of time as consultants to companies and governmental departments.

2. I have personally worked on contracts where it was quite clear that the client wanted to train some of its employees to work in this area and the project I was engaged on was a kind of on-the-job training for them.

actively seeking to acquire the necessary skills in this way, client companies will have little option other than hiring consultants on a frequent basis.

OPTION 3—RECRUIT EXISTING EXPERTS

When a company commits to design for accessibility, it might make financial and competitive sense to bring the expertise in-house on a long-term basis. One possible method for achieving this is to find and recruit existing accessibility practitioners. However, there are not many of these practitioners out there, so this resource will, in a very short space of time, become scarce. Another option is to recruit from other sources. For example, a number of universities are building strong research teams in the areas of Universal Design and Universal Access. If a company seeks to acquire the skills necessary to design for accessibility, a short-term solution would be to employ to members of those university research teams.

Unfortunately, though, these solutions are not sustainable in the longer term. There are a great many companies who need to acquire workers skilled in this area and not enough such people to go around. It takes a long time to train up new people with those skills. For instance, looking at the university route, most researchers typically have to complete Bachelors, Masters and PhD degrees before being considered "expert." Even then it is likely that some level of post-doctoral experience would be preferable. In addition, there is a difference between performing research in an academic environment and designing for accessibility in a business context. Specialists recruited from universities would still need time to familiarize themselves with working in industry before they could become fully productive members of a company's design team.

The law of supply and demand would appear to dictate that these sources of skilled workers will become increasingly scarce following a combination of repeated plundering together with the comparatively long time taken to train people to the required level of expertise, which will most likely result in the price of the resources (i.e., salaries) increasing.

OPTION 4—DEVELOP PROFESSIONAL TRAINING

Perhaps the most sustainable option for a company is simply to train new or existing staff in the skills required to design for accessibility. This way the company would not be solely dependent on recruiting or hiring external experts.

Education

The most suitable internal candidates to be trained in design for accessibility are the company's own designers and usability practitioners, as they primarily only need to learn an additional skill set that complements their existing knowledge. However, this solution is not as straightforward to implement as one might hope. Again, the absence of a recognized national or international accreditation body that vets and approves or endorses particular educational training courses makes it difficult to know which courses are worth investing in. Currently, perhaps the most reliable "brands" that can be relied on are courses from established, renowned universities. However, these courses are not easy to find and places on them fill up quickly.

Training courses are not the only option, though. Other educational materials (this book, for instance) are well worth acquiring. Since this is a comparatively new field, research papers offer a valuable information resource. At the time of writing, the ACM Digital Library has 3480 peer-reviewed papers focused on "accessibility" (ACM, 2005). Searches on terms such as "disability" and "impairment" yield additional results.

Participation in Design Competitions

Design competitions are also a great way of familiarizing designers with the concept of accessibility. For example, the Helen Hamlyn Research Centre at the Royal College of Art in London runs their annual Design Challenge, sponsored by the Design Business Association (DBA). In this competition, six teams are paired up with groups of "critical" users (typically users with functional impairments, such as low vision or limited mobility) and challenged to develop new products that are responsive to the needs of those users (Cassim, 2004). This format has proven to be both very effective in helping designers become aware of the needs and aspirations of the users that they worked with. The competition is also very popular with the participants.

Along similar lines, the Association of Computing Machinery (ACM) Special Interest Group on Computers and Accessibility (SIGACCESS) has recently launched an annual Student Research Competition for graduate and undergraduate students.

Exemplar Case Studies

Organizations such as the North Carolina State University, the Trace Center at the University of Wisconsin-Madison, and the UK Design Council host Web sites that list design best practice in this area, as well as often providing interesting case studies (see Chapter 1). Turning the case study idea around, Web sites are now appearing that offer examples of general usability issues where designs are found to be somewhat lacking (Hurst, 2005). These often-humorous sites offer a great way of getting designers thinking more carefully about what they are doing. A classic example is that of a sign for a restroom in a Japanese airport on which the Braille has been painted (rather than embossed) and, to make matters worse, positioned some three meters above the ground.

COMBINING THE OPTIONS

The options discussed above are not mutually exclusive. They may be implemented in varying combinations. There are potentially interesting business models based on combinations of external expertise and computer-based tools. As discussed earlier, one of the drawbacks of computer-based tools is that they frequently require expert set-up and expert analyses of results.

Imagine that a fictional budding web-design company, ABC Corp., decides to adopt an equally fictional computer-based tool, ArachnoidAccess (aAccess for short), which assists in the design of accessible Web sites. aAccess provides prompted guidance through the process of designing a web-page (e.g., "Remember to include a breadcrumb trail, so users know where they are in the site") and also offers a "Check Accessibility" feature that lists any broken (i.e., incorrect or incomplete) HTML code and violations of the Web Accessibility Initiative guidelines (W3C, 2005). Initially, the Web site designers need assistance in configuring aAccess to meet their needs and learning all of its features and functions. They decide to call in a team of (fictional) expert consultants, ArachnoidConsultants, Inc. (aConsultants), the company that developed aAccess. aConsultants set-up the software and provide initial training on how to use it. This training consists not only of a traditional software-training course overview of the functionality of the software, but real case studies showing the software in use on real projects.

Once the period of initial training is complete, the designers at ABC Corp begin using the software on their customers' sites. However, because they are still new to this field, they do not want to break away completely from the support offered by aConsultants. Instead they agree to a second-level monitoring contract. In this new contract, aConsultants are able to monitor the output of the ABC Corp designers remotely through functionality built in to aAccess. They are able to provide feedback and assistance when aAccess flags a particular problem in a Web site's design or code. In this model, ABC Corp has support from aConsultants when it is needed, but otherwise is free to pursue its business as it sees fit. This can be thought of as just-in-time access to expert knowledge. Similarly, the experts working at aConsultants are able to offer their expertise to the maximum number of clients through their remote-monitoring capabilities. Such a model maximizes the utilization of the limited number of domain experts.

IDENTIFYING THE WIDER SETS OF SKILLS NEEDED

Clearly, any company looking to implement design for accessibility needs to acquire skills in all relevant accessibility tools, techniques and practices, most likely through the options discussed in this chapter. However, that is most likely only part (albeit a very important part) of the skills that are needed. Creating a truly accessible and usable product is only possible if all of the other aspects of the product design are of a sufficient quality to support them. For instance, look at the design process for creating a Web site that is accessible to someone who is blind. A basic solution to meet that user's needs is to ensure that, for example:

- keyboard shortcuts are available and implemented correctly to replace the usual mouse input, such as clicking on a hyperlink;

- no information is provided without a textual equivalent; and,
- the site is compatible with screen-reader software that can provide a voice output version of each page.

A full comprehensive set of requirements would be much longer than this—see the Web Content Authoring Guidelines (W3C, 2005) for a more complete treatment of this issue.

While these solution steps can be readily identified by an experienced design for accessibility practitioner, implementing them requires a high-level of web-coding expertise, often beyond the level found in the majority of web pages. Unlike programming in a language such as C or C++, where applications will not compile and run of the code is not completely correct, web browsers, such as Firefox and Internet Explorer, have been programmed to display web pages even with "broken" HTML coding. In effect, they take a best guess at what the coding was meant to be. Often it does not matter if the rendering of the page is not quite what the page designer intended, because as long as all of the content is visible somewhere on the screen, the user should be able to find the parts that are of interest.

However, screen-readers are much more limited in their functionality and intelligence than a human reader with sufficiently good eyesight. They can only read a page in a linear fashion and can only identify important items, such as links, if they have been coded and tagged properly. If a site has not been coded correctly and has "broken" code, the screen-readers may well be unable to operate successfully on that site.

Thus, it is clear that to design and implement an accessible Web site requires not just knowledge of the users and the functionality that needs to be added to a Web site to meet their needs, but also the programming skills to ensure that the site has been coded correctly throughout its entirety. Designing for accessibility therefore requires that companies be expert in, and have full control over, all of the attributes of their products. This should be straightforward for a company that makes many of its products from scratch. However, a company that assembles many of its products from off-the-shelf components may find it more difficult to implement design for accessibility on its own. In this circumstance, it is necessary to put a framework in place for bringing the component suppliers on board and work with them to encourage the adoption of design for accessibility in *their* products. This example shows that implementing design for accessibility requires more than just hiring an "accessibility" expert. It may require a wide range of skills, from coding to managing a new type of relationship with suppliers. Companies looking to implement design for accessibility need to think in terms of the big, strategic picture as well as the smaller tactical issues.

KEY POINTS

- Expertise in design for accessibility and inclusive design is scarce at the moment.
- Companies need to implement a proactive policy to ensure that they can acquire the skills that they need to meet their own design for accessibility targets and comply with legal regulations.
- This chapter proposes four possible methods for companies to acquire those skills.

7

Case Study—Making
Expertise Available
Within a Company

The preceding chapters have discussed the broad actions that a company should take to implement design for accessibility into its product design processes. Central to that is addressing the challenge of acquiring the necessary skills within the company and then making them available to the design and marketing teams at the appropriate time. The purpose of this chapter is to discuss how one company, IBM, is responding to this challenge. The aim here is not to promote this particular solution as *the* solution, simply to highlight one *possible* solution.

IBM needs little introduction. It is a large multinational corporation with interests covering all aspects of information technology. Perhaps less well known outside of business circles is that IBM derives a significant proportion of its earnings from what would be considered traditional consultancy, business process transformation, outsourcing and the like. As such, IBM has a multitude of teams developing a very wide range of product lines, encompassing hardware, software, and knowledge-based products. Many of those teams are highly specialized and know a great deal about very focused product areas. Other teams specialize in much more transferable skill-sets, developing generic processes and products that can be applied on a variety of situations.

Within IBM all major product development efforts employ a formalized user-centered design process that integrates user input and user interface evaluation and design methods into the development process, and user and customer satisfaction is tracked over product releases and against competition. However, including users with disabilities in this process has been, with a few notable exceptions, rare.

Until recently, the practice of making products easy to use has focused on the average, or the most common, or most influential user of the product being designed. This has translated into testing and designing products for the most common human characteristics and thus excluding users with most disabilities. Even typical user-centered design selection methods (representative sampling of the user population) work against including users with disabilities because of the large under representation of persons with disabilities in the work force. In response, IBM is undertaking to include persons with disabilities in its product design and development process, to define sets of best practices for doing so, and to make its offerings easy to use for users with disabilities.

AN INTRODUCTION TO ACCESSIBILITY WITHIN IBM

IBM has a long history of interest in design for accessibility. In an effort to bring the expertise together into a single point of contact, IBM established its worldwide Accessibility Center in 1999 (IBM, 2005a). The primary mission of the Accessibility Center was to ensure that all of IBM's products and services are accessible and comply with all relevant legislative requirements, principally those defined within Section 508 (WIA, 1998). The Accessibility Center also addressed the need to ensure that IBM's internal IT tools and work environment are accessible for employees with disabilities. The Accessibility Center has recently expanded its remit and is now the Human Ability and Accessibility Center (HA&AC).

It has become clear that, as IBM's experience with accessibility continues to develop and evolve, making products accessible does not necessarily mean they are easy to use for users with disabilities. In many cases, the concept of access means just that: users with disabilities are to be provided with access to the utility, i.e., function and information, offered by a product. The reason for this lies in the framing of the legislation and most accessibility standards, such as the WAI guidelines (W3C, 2005).

To repeat an example given earlier in this book, it is easy to define a requirement that states that ALT TEXT be present for all images on a Web site. It is equally easy to test whether such a requirement has been met. It is more difficult to specify a requirement that the ALT TEXT set make sense (how is "make sense" defined?) and much more difficult still to test that the requirement has been met. Various development organizations within IBM learned this over the past few years and had begun to address this issue independently of each other. It quickly became apparent at a corporate level that a more strategic and centralized model was required to plan a structured and cogent approach to looking beyond mere "access."

ACCESSIBILITY GUIDANCE WITHIN IBM

As with most other companies in the IT industry, IBM's accessibility practices have been influenced heavily by recent legislation, such as Section 508 (WIA, 1998) and the Americans with Disabilities Act (ADA, 1990). Prior to 2004, IBM's accessibility efforts were focused on meeting accessibility standards which enable its products to work with assistive technologies used by people with disabilities. IBM recognizes that accessibility goes beyond issues of compliance and the adoption of a holistic approach is required. To this end, a four-tier framework has been adopted (Keates, 2005)—see Figure 7-1.

From the framework it can be seen that the current level of industry's response to accessibility is focused squarely on compliance (Tier 1). While meeting compliance requirements is of some use to end-users, the real benefit to the overall user experience comes when industry looks beyond compliance towards a person's total quality of life.

Implementing this framework begins with implementing accessible technology infrastructure and ends with business transformation.

Tier 2 (Experience Driven) looks at developing technology and products that aim to assist the user in completing their stated task. It is the next logical step to look at beyond simple compliance and mere access. Thus the questions asked here would not be of the "Can you access each item on this menu?" sort, but rather "Can you compose and send an e-mail with this program?" It is about ensuring access to the complete application, not just each of its components.

Tier 3 (Relationship Driven) takes a more holistic view. Now the question has become "What range of applications do you need for complete access to the information society?" The idea here is that applications should not be considered in isolation, but rather as part of a whole relationship between the company and its clients.

Finally, Tier 4 (Societal Transformation) looks at structural changes to the environment in which the user operates. These are primarily background changes in the provision of services to the user. The idea underlying these changes is that once the user has a view of the information society that he or she is comfortable and happy with, the onus is on the organizations supplying services to the user to comply with the user's view of the world, rather than the other way round. Most technology available today takes the diametrically opposed view that the user should adapt to the service.

Each of the four tiers represents an expanding solution domain and the model could just as easily be presented as four concentric circles, with Tier 1 in the center and Tier 4 as the outer circle.

TAKING ACCESSIBILITY FORWARD WITHIN IBM

While the model is straightforward to describe, implementing it represents significant challenges. Even Tier 1—compliance driven—is comparatively poorly defined at the moment. Until the courts decide what constitutes "reasonable accommodation" reliably, then the best a company can hope for is to cover as many bases as possible, as well as possible. Even then, there is no guarantee that a judge may not disagree that the company has taken all reasonable steps to include particular user groups.

The approach taken to Tier 1 within IBM is to define a cross-company set of accessibility requirements that each product and service offering needs to meet.

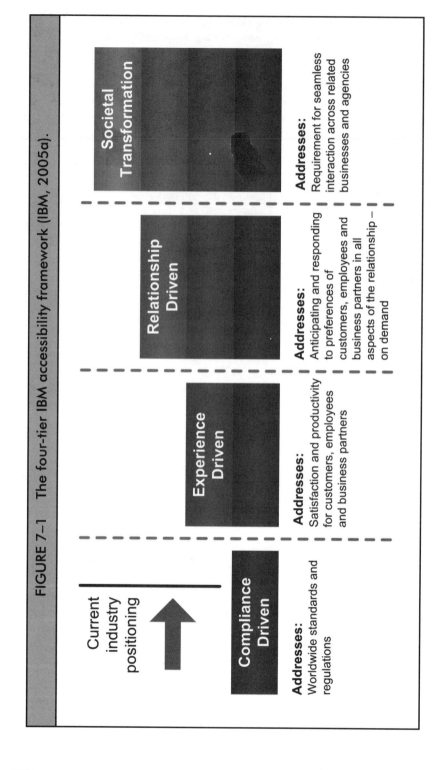

FIGURE 7-1   The four-tier IBM accessibility framework (IBM, 2005a).

Societal Transformation

Addresses:
Requirement for seamless interaction across related businesses and agencies

Relationship Driven

Addresses:
Anticipating and responding to preferences of customers, employees and business partners in all aspects of the relationship – on demand

Experience Driven

Addresses:
Satisfaction and productivity for customers, employees and business partners

Current industry positioning

Compliance Driven

Addresses:
Worldwide standards and regulations

These checklists largely reduce the process of design for accessibility to "Is *this* product accessible?" If it is, then the product passes that particular check, if not then it needs to be revised. The checklists cover the following topics and are available from the Human Ability and Accessibility Center (IBM, 2005b):

1. software accessibility;
2. web accessibility;
3. Java™ accessibility;
4. IBM Lotus Notes accessibility;
5. hardware accessibility;
6. peripherals accessibility; and,
7. documentation accessibility.

This checklist approach is a direct consequence of how the standards and legislation governing accessibility have been structured and phrased. It is generally recognized within IBM that the checklist approach is a starting-point and not the end-point, when it comes to designing for accessibility. So, the question then arises of how to take design for accessibility forward, to Tiers 2 through 4?

IBM has started to deal with this issue by looking beyond simple utility (functionality) and access, and instead at the overlap between accessibility and usability. The company has recognized that both of these issues need to be addressed to ensure that users get the best possible user experience. To acknowledge that this is a new twist on the existing accessibility and ease of use ideas present within the company, a new name was adopted for this approach—Usable Access.

WHAT IS "USABLE ACCESS"?

Usable Access is a joint initiative of the accessibility and ease of use communities in IBM, chartered to focus on improving ease of use for people with disabilities. It is intended to bring together best practices from the fields of accessibility and usability into a single framework for implementing the goal of Usable Access. to address this issue, the Usable Access work group was formed in 2004 as a cross-IBM team representing many areas in IBM that have been involved in accessibility and ease of use. Members of the team were drawn from groups such as the Human Ability and Accessibility Center, Accessibility Research, Software Group, and hardware divisions.

The goal of the work group was to develop Usable Access best practices that could be deployed by IBM development teams. It was expected that those Usable Access best practices would be used to expand the focus of User Centered Design and User Engineering methods to be inclusive of all users. Thus, this would expand the focus of accessibility from product interoperability with assistive technology products to include "ease of use." While work had been done in some academic and research labs on integrating usability and accessibility practices, this focus on Usable Access was considered relatively new within commercial product development organizations.

USABLE ACCESS BEST PRACTICES

The root cause of industry producing products that are difficult to access even when they are accessible is twofold. First, people with disabilities are not normally included in the product user-centered design process and the typical user interface designer does not understand the unique characteristics of people with disabilities. Second, the general approach to making products accessible pose challenges to designing ease of use for people with disabilities.

Many interfaces are designed using one input mode, typically motor, and one output mode, typically visual. The reason for this is that the use of multiple modes of input or output are generally considered additional extravagances, not normally needed to achieve a particular task. That is, of course, assuming that the user has the necessary physical capabilities to communicate with the interface through the pre-scribed modes. Should the user not have the requisite capabilities, then the interface becomes inaccessible and, consequently, unusable. The interface then needs either re-designing to support other communication modes, or else retrofitting with some kind of bolt-on technology to convert one mode into another.

Unfortunately, though, product designs that are predicated on the use of a single mode of input or output capabilities are difficult to retrofit for easy use by users with different sensory and manual capabilities. For example, accessing graphical user interfaces (that demand primarily visual capability to operate) can only be achieved through the use of a screen-reader for someone who is blind. The resultant interaction is sub-optimal for such users because information that has been optimized for one perceptual domain (that of sight) has to be transformed into a different perceptual domain (that of hearing). Hearing is a linear mode of interaction, whereas sight is a two- or three-dimensional one, depending on the nature of the interface. Thus the screen-reader needs to either somehow represent the additional dimensional information to the user, or else ignore it. If it ignores the additional information, then the user loses potentially useful information. On the other hand, if it tries to interpret the second and third dimensions (where appropriate), it needs to do so in a way that is reliable, repeatable and also makes sense, without overwhelming the user with unimportant information. To illustrate this challenge, consider two examples.

In the first example, the user is trying to select an icon on the desktop. The location of the icon and how it is positioned relative to other icons is most likely unimportant, what matters is the link that it represents, e.g., launching a word processor or e-mail program. The second example is a data item in a table. Here the value and position of the data item need to be considered, i.e., it matters which column it is in and also how its value compares with those around it.

Therefore, a global set of rules or heuristics about how to interpret and transform two-dimensional visual interfaces into one-dimensional aural ones needs to be able to recognize what the underlying purpose of the interface is and try to select the most appropriate rules for the transformation. This can be achieved to a more or less satisfactory level for something as identifiable and well understood as a data table, but is considerably more difficult for, say, a label on a diagram. In this case, the screen-reader would need to be able to recognize that a diagram exists, interpret what

CHAPTER SEVEN

the diagram represents, recognize and interpret that a text label exists, and finally interpret how the spatial location of the label relates to the underlying diagram. Such capability is beyond screen-readers and will most likely remain so for the foreseeable future. Problems like this are avoided on web pages by using additional text labels such as ALT TEXT to describe the diagram and remove the need for the screen-reader to perform any interpretation of the image. However, such mark-up is not commonly used on other parts of the typical graphical user interface. Consequently, the aim of the Usable Access workgroup was to provide guidance to designers and product developers to help them avoid such pitfalls.

OUTPUT OF THE WORK GROUP

After an initial period of consultation with accessibility experts across the company, it was decided to document the available best practice information about the following aspects of designing for accessibility:

- performing product evaluations with users with disabilities;
- user interface design and implementation techniques for people with disabilities;
- use of personae of users with disabilities;
- complex visualization applications;
- hardware enablement issues; and,
- lessons learned from developing IBM assistive technologies.

The first two areas represent a great deal of experience across IBM, industry, and academia that could be utilized and documented. The use of personae is a user-centered design technique in IBM that could easily be adapted for use with people with disabilities.

Complex visualization applications, such as visual editors, are an increasingly important technique used in the industry to provide significant enhancements to a user's ability to create new applications and deal with very large amounts of data. Visually impaired users, however, not only cannot take advantage of these enhancements, they are also prevented from participating in work where teams of fellow workers use these visual tools to do their work. Therefore, the Work Group decided to initiate work to understand how to make such applications fully accessible and easy to use.

Much of the Usable Access work described above focused on software products, and so the work group initiated a focus on unique hardware product issues. Finally, IBM has a long history of developing assistive technologies, such as IBM Home Page Reader and the IBM Java Self-Voicing Development Kit. Through the development of these technologies and through supporting interoperability with individual product teams, a number of important lessons have been learned about making product user interfaces easier to use through assistive technologies.

Sub-groups of topic specialists were formed to address each of the areas identified above (user evaluations, use of personae, etc.). Each sub-group in turn produced a white paper (e.g., IBM, 2005c) that summarized the state-of-the art of existing best practice and also how such practice should be applied with the product development process within IBM. Where appropriate and available, specific examples of the best practice in use were given. For example, sample personae were described in the personae paper. The white papers produced were reviewed at both the sub-group and work group levels and revised iteratively. In late 2004, the white papers were then made available to other groups within IBM as part of a wider consultative phase to determine whether product development teams were able to use them in the format provided.

INSTITUTIONALIZING USABLE ACCESS

The IBM Human Ability and Accessibility Center has developed a strategic framework which views the development of this confluence of accessibility and ease of use as a natural evolution of the IT industry's focus on accessibility. Usable Access is no longer just a buzz-phrase. The working group, whose remit was initially to collect and collate best practices in this area has now evolved into the Usable Access Council. The membership of the Council is fundamentally similar to that of the working group, i.e., accessibility, usability, and inclusive design experts from across IBM's divisions and business groups. The biggest difference between the working group and the new Council is that the Council is now a recognized body within the corporate hierarchy. It exists not only to gather and update the available documentation on best practices, but also to provide practical guidance and individual expertise to project teams, typically through a member of the Council being nominated to lead the contact with the project team. The Council also keeps records of employees with disabilities and also agreements with external user groups so project teams can gain access to such users in the shortest possible time.

In summary, the Council represents a one-stop-shop collection of subject-matter experts and users, who can be contacted and consulted by project teams across all of IBM. The Council itself has a mandate to keep investigating new and developing practices and technologies in this area, to ensure that the subject-matter experts are up-to-date with the very latest information and techniques.

SUMMARY OF THE USABLE ACCESS INITIATIVE

Despite an increased focus on the accessibility of Information Technology offerings, IT products are typically not easy to use for users with disabilities, even when they are "accessible." One of the major root causes for accessible, but difficult to use, products is, simply put, ignorance of the requirements and characteristics of people with disabilities.

IBM has undertaken efforts to change this, by defining a set of relevant best practices and integrating a focus on users with disabilities into its product design and development processes. This issue is being addressed by providing guidelines for incorporating users with disabilities into usability and user-centered design evaluations, by providing guidance and a set of templates for developing personas for users with disabilities, and by providing a set of user interface design and implementation guidelines.

Obtaining feedback from users about product design through sessions that allow current and prospective users to exercise early product prototypes is essential when trying to make products easy to use. While testing individual products is critical, translating feedback into useful designs will be enhanced by the user interface design guidelines and techniques that have been developed under this initiative. As user interface designers employ these guidelines, and temper their designs with the input from usability evaluations with users with disabilities, IBM will gain the additional knowledge needed to improve and extend the current set of guidelines.

This is just the beginning of this work and it is expected that there will be a period of intensive learning over the next few years where the guidance needs to be improved. Support mechanisms will be established with the dual purpose of helping IBM user experience and accessibility professionals employ these guidelines, and collecting the data needed to make continuous improvements. As part of this, areas will be identified to improve deployment, particularly in the area of tools, and to develop more advanced best practices in usable access.

As the IBM internal methods improve and products come to embody Usable Access principles, it is expected that IBM's ability to provide not only exemplary product user experience, but also consulting services to its full range of customers, will be enhanced. The potential impact on the lives of persons with disabilities is significant.

LESSONS FOR OTHER COMPANIES

As stated at the start of this chapter, this is not intended to sing the praises of IBM. It is simply a presentation of one approach that one company is taking. Having said that, it is an approach that has been thought through carefully and supported by the highest levels of the corporation. It has also been resourced well. Access to all the necessary expertise and information within the company has been secured. The resultant documentation on best practices represents almost all the relevant information needed by project teams within IBM. As such, the risk of individual project teams trying to re-invent the accessibility wheel for each new project has been minimized. This ensures that development costs arising from accessibility requirements are kept to a minimum—the teams know where to go to get the information they need as quickly as possible. The dialogue between the design teams and the Usable Access Council, and the living nature of the best practices documentation, ensures that the Council is able to update and revise its advice based on what has been shown to work within the company and what has not.

There are several important lessons for other companies. For example, initiatives of this sort need to be vocally supported by senior management, otherwise it is very difficult to secure access to all the necessary people across the company, many of whom are very busy and whose time is in high demand. Additionally, it is very important that the data gathered about best practices be as transferable as possible. If it is not transferable, then expertise needs to be made available to help project teams interpret the guidance.

Such initiatives are not only useful internally for improvement in design for accessibility practices, but also for external public relations. A company that actively implements and supports such initiatives gives itself, as a result, the opportunity to enhance its public image. After all, the company is saying, in effect, that it really cares about all of its customers. However, as with all public relations exercises, it must be remembered that any such exercise must be planned and coordinated carefully to ensure that all resultant publicity is in line with the company's brand image.

KEY POINTS

- The Usable Access initiative was actively supported by IBM's senior management.
- The collected best practices are available across all of IBM via the intranet and are backed by a network of accessibility experts.
- The collections of best practices are living documents—continually being reviewed and updated to ensure their continued relevance and applicability.

8

Putting Accessibility Into the Design Process

In the previous chapters, we have examined how the different levels of company management can promote a culture that supports and nurtures design for accessibility practices. In the next few chapters we will examine the impact of designing for accessibility on the design practitioners, the people who actually have to design the products.

A USER-CENTERED DESIGN PROCESS

Designing for accessibility requires that the user be placed at the center of the design process. Making assumptions about user behaviors and characteristics will simply not be good enough. All the decisions made during the design process must be based on reliable information about what the users genuinely need and want from the product. This stipulation may sound draconian, but it could easily be argued that all good design practices are based on this. It does, however, raise the issue of how to handle all the information about the users.

The Need to Manage User Information

Providing designers with either insufficient or too much information about the users can impair their effectiveness in developing accessible products or services successfully. The current state in most companies is that the data needs of designers are being met by them trying to estimate (or even guess) information about the users' capabilities. This is clearly an unreliable approach and prone to error.

It is tempting to imagine that simply providing designers with more and more data about the users will be sufficient to ensure an accessible product is designed. There are several data sources readily available about people and their capabilities, for example standard human factors texts about human capabilities. These texts are full of general data about body-part sizes and strength characteristics. For example, Adultdata has 266 body dimensions and 28 strength measurements (Peebles and Norris, 1998). This data is remarkably useful in particular situations, but it is fair to say that most designers would struggle to use it when designing for accessibility.

The reason for this is that most common human factors and anthropometric texts focus on providing information at the population level (e.g., Peebles and Norris, 1998). However, much of this information focuses on the 5th to 95th percentile ranges. When considering design for accessibility, it is often those users outside of this range that are of interest. Of course, the methods of providing population data can also be replicated for subsections of the population, as the underlying data formats are still valid. However, the resultant data collections still end up being a similar size to those for the complete population. For example, anthropometric texts describing the characteristics of older adults (e.g., Smith, Norris, and Peebles, 2000), have a similar number of tables and charts as those for all adults (e.g., Peebles and Norris, 1998). This makes sense, because there are similar numbers of body attributes to report. Thus, while texts such as Older Adultdata help the designers and researchers obtain more specific information about older adults, the overall problem of sheer volume of data is not reduced.

Consequently, there is a real danger that unless the information provided is carefully screened to ensure that it is relevant to the use of the product, the designer may be overwhelmed by the sheer volume of data that could be provided about the end-users. The ideal solution for designers is to have sufficient data available to hand when it is needed—a kind of data "just in time." The need for sufficiency is central to this approach—the data needs to contain all the necessary information, but no more, to avoid information overload.

In summary, it is essential to identify the information about the users that is of most use to designers and then to find ways of obtaining that data and presenting it in an easily digested format. As would be expected, given the wide variety of possible information sources about users and the different data requirements of designers, there is no single ideal approach for all circumstances. However, there are techniques available that enable the necessary information to be identified, gathered, and packaged to meet the needs of designers.

For example, Quality Function Deployment (QFD—Akao, 1990) is a design methodology that has been developed specifically to improve the number of customer compliments about a product. This popular methodology represents an evolution of the principles of Total Quality Management (TQM—Dahlgaard, Kristensen, and Khanji, 2005). QFD came about because of a recognition that the best result traditional quality techniques can expect is zero faults. This result may be sufficient to satisfy the goal of practical acceptability, but contributes very little towards a product's social acceptability. QFD corrects this imbalance by including measures that focus explicitly on the users' needs and wants—specifically minimizing customer dissatisfaction with the product (Zultner, 1993). The ultimate aim of approaches such as QFD is to support the designer in developing products that are both socially and practically acceptable, through developing the necessary understanding of the users to accommodate their wants and needs proactively during the design process and obviating the need for retrospective adaptations. In other words, information and data about the users are vital to the success of design for accessibility.

THE 7-LEVEL DESIGN MODEL

The 7-level model is an example of a design approach that has been structured specifically to meet the challenges of designing for accessibility. In this section, we will examine how it was developed.

When designing any new product, it is necessary to adopt a three-stage development strategy (Blessing, Chakrabati, and Wallace, 1995):

- **Stage 1**—define the problem—gain an understanding of the system requirements;
- **Stage 2**—develop a solution—develop a system that includes consideration of able-bodied and motor-impaired users;
- **Stage 3**—evaluate the solution—make sure that the solution is effective.

These stages are also applicable for designing an interface. The inclusion of a broader range of user capabilities affects all three of the above stages:

- **Stage 1**—the problem definition should explicitly include reference to the intended target users;
- **Stage 2**—an appropriate design approach should be adopted for the target users;
- **Stage 3**—the target users should be included in the evaluation process.

In order to produce a usable and accessible product or service, it is necessary to adopt strongly user-centered design practices. It is important to be able to modify and refine the interface iteratively, combining both design steps and usability evaluations, which typically involve measurement against known performance criteria. Nielsen gives an account of the use of such criteria in a method known as heuristic evaluation (Nielsen, 1993). Developing a usable product or service interface for a wider range of user capabilities involves understanding the fundamental nature of the interaction. Typical interaction with an interface consists of the user perceiving an output from the product, deciding a course of action and then implementing the response. These steps can be explicitly identified as perception, cognition and motor actions (Card, Moran, and Newell, 1983) and relate directly to the user's sensory, cognitive and motor capabilities respectively. Three of Nielsen's heuristics explicitly address these functions:

- *Visibility of system status*—the user must be given sufficient feedback to gain a clear understanding of the current state of the complete system;
- *Match between system and real world*—the system must accurately follow the user's intentions;
- *User control and freedom*—the user must be given suitably intuitive and versatile controls for clear and succinct communication of intent.

Each of these heuristics effectively addresses the perceptual, cognitive, and motor functions of the user. Building on these heuristics, a design approach has been developed that expands the second stage of the design process, solution development, into three specified steps (Keates and Clarkson, 2003). Each level of the resultant design approach, shown in Figure 8–1, is accompanied by user trials throughout and a final evaluation period before progression to the next level, thus providing a framework with clearly defined goals for system usability.

The 7-level approach addresses each of the system acceptability goals identified by Nielsen (1993). The approach has been applied to a number of case studies including the design of a software interface for an interactive robot (see Chapter 9) and the review of an information point that shall be discussed in more detail later in this chapter.

Applying the 7-Level Approach

The 7-level approach has been structured as a high-level model, so each level represents an aim, but the method of achieving that aim can vary according to the expertise and knowledge of the designer.

- Level 1 identifies the user needs, i.e., the social motivation for designing the product. This can be identified through softer, sociological assessment methods. Questionnaires and interviews are good methods for identifying the user needs.
- Level 2 focuses on specifying the required utility of the product. Traditional engineering requirements capture techniques (Beitz and Kuttner, 1994) can be used, as can task analysis (Nielsen, 1993; Card, Moran, and Newell, 1983). Alternatively, functional assessments of rival products or observation of existing methods can provide insight into the necessary functionality.

Levels 3 to 5 focus on the stages of interaction. Usability and accessibility techniques can be applied directly to these levels, as can anthropometric and ergonomic data and standards. Prototypes of varying fidelity play a key role in these levels.

- Level 3 addresses how the user perceives information from the system. This involves assessing the nature and adjustability of the media used, their appropriateness for the utility, and the physical layout. Anthropometric data are important to ensure that the output is in a position that the user can perceive it. Ergonomic and empirical data from trials are also necessary to ensure that the stimuli are intense enough to be perceived. Ideally, environmental conditions, such as lighting and noise, also need to be identified and modeled.
- Level 4 assesses the matching of the system contents and behavior to the user mental model of the system. Once the output channels are defined, the content/utility can be added to the system and evaluated because the functionality for monitoring the system is in place. Literally, the user can see/hear/etc. the data. Common techniques to map the user system behavior to user expectations include cognitive walkthroughs.
- Level 5 focuses on the user input to the system. As with level 3, this involves assessing the nature and adjustability of the media, their appropriateness for the utility, and the

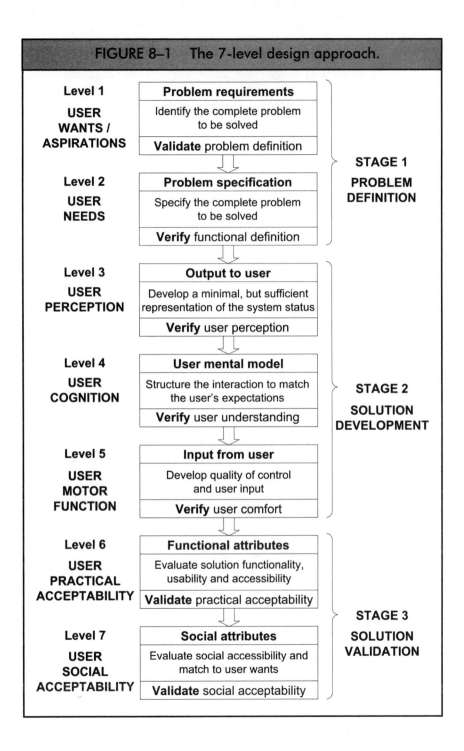

FIGURE 8–1   The 7-level design approach.

Level 1 **USER** **WANTS /** **ASPIRATIONS**	**Problem requirements** Identify the complete problem to be solved **Validate** problem definition	**STAGE 1** **PROBLEM** **DEFINITION**
Level 2 **USER** **NEEDS**	**Problem specification** Specify the complete problem to be solved **Verify** functional definition	
Level 3 **USER** **PERCEPTION**	**Output to user** Develop a minimal, but sufficient representation of the system status **Verify** user perception	**STAGE 2** **SOLUTION** **DEVELOPMENT**
Level 4 **USER** **COGNITION**	**User mental model** Structure the interaction to match the user's expectations **Verify** user understanding	
Level 5 **USER** **MOTOR** **FUNCTION**	**Input from user** Develop quality of control and user input **Verify** user comfort	
Level 6 **USER** **PRACTICAL** **ACCEPTABILITY**	**Functional attributes** Evaluate solution functionality, usability and accessibility **Validate** practical acceptability	**STAGE 3** **SOLUTION** **VALIDATION**
Level 7 **USER** **SOCIAL** **ACCEPTABILITY**	**Social attributes** Evaluate social accessibility and match to user wants **Validate** social acceptability	

physical layout. Again anthropometric measures are important to ensure that the input media are within the operating range of the user. Ideally, empirical data from user trials needs to be gathered to evaluate the effectiveness of the input solutions. These can be supported by adopting user-modeling techniques. Where user trials are impossible, suitably calibrated user models can be used to provide design data.

- Level 6 involves the evaluation of the complete system to ensure satisfactory utility, usability, and accessibility. Formal user trials and usability/accessibility assessments are essential at this point, before the design can progress to the final level, 7.

- Level 7 assesses the resultant system against the user needs. This mirrors Nielsen's social acceptability requirement. Softer, more qualitative approaches are generally needed, such as surveys, interviews, and questionnaires.

Although the 7-level approach is presented as a flow diagram, in practice each of the stages can be applied in an iterative manner. Indeed, in many circumstances, it may prove essential to iterate within and between levels. However, the general order of levels should be applied in the order shown in Figure 8–1.

IDENTIFYING WHAT USERS WANT AND NEED

Many designers are not trained to identify what the user wants and needs. In the case of designing for accessibility, the user wants and needs include additional factors beyond those usually considered for so-called mainstream products, such as the effects of age, experience, and impairments. As this information can be quite complex, it is obviously difficult to identify and provide exactly the right data to the designer. Common approaches often focus on providing inspiration to the designers, encouraging the right frame of mind, rather than a list of specific requirements. For example, the provision of anecdotes or vignettes about the life of a target user is a popular method, and contributes to the "design by story telling" approach adopted by companies such as IDEO (Nussbaum, 2004). Another approach is simply to allow the designers to spend time with end-users, to enable them to become familiar with the users, their capabilities, needs and so on. This is the approach used for empathic design (Gheerawo and Lebbon, 2002).

Traditionally, defining the user expectations for a product has been the remit of market researchers, or other specialist professionals, such as ethnographers. This does not necessarily need to change, as long as those professionals can provide sufficient information about the user wants and needs to the designers in a format that they can understand and use. The following represents a summary of typical methods used by researchers to elicit the user wants.

- questionnaires
- interviews
- user observation sessions/user trials
- focus groups
- ethnographic methods

Each of the above methods can be effective in eliciting the user wants, but they also each have their potential limitations.

The questions asked on questionnaires have to be prepared before any data is actually obtained from the users. The questions are typically open or closed. Open questions allow for more detail from the respondents, but are difficult to analyze statistically because there is no guarantee of "quantifiable" data being gathered. Closed questions, that is questions that ask for a yes/no type answer or else answers on a sliding Likert-type scale from 1 to 7, say, are well suited to statistical analysis, but will only gather the data related to the questions asked. As such, it is possible that potentially useful data on other topics not asked about is not gathered. In effect, questionnaires are only as useful as the quality of questions asked allows.

Irrespective of the quality of questions on the questionnaire, if they are mailed out to respondents (as compared to being completed in the presence of a researcher), then they also often suffer from low response rates. The advantage of questionnaires, though, is that they can be mailed out to lots of people and that goes some way to compensating for the expected low response rate. However, not all respondents are equally likely to reply (or not). It is quite probable that some underlying factor may make particular respondents more likely to answer than others and that factor needs to be identified and isolated to make sure that the data gathered is not, in some way, inherently biased. For example, if conducting a survey on attitudes to accessibility in companies through the use of a mailed out questionnaire, it is quite obvious that some companies will be more willing to discuss their attitudes than others. Companies with proactive policies promoting design for accessibility among their design teams will often be very happy to spread the word of what good corporate citizens they are. On the other hand, companies who do not have any particular policy towards accessibility will most likely not want to reveal that fact. Thus the responses to the questionnaires will most likely be skewed into over-representing the number of companies that actively address accessibility.

Interviews can be structured, semi-structured, or unstructured. Unstructured interviews are completely freeform and offer maximum flexibility for obtaining information from respondents. The downside is that the data collected is likely to vary considerably from interviewee to interviewee (and also from interviewer to interviewer) and so it is difficult to draw statistically significant generalizations from the data gathered. In effect, unstructured interviews can be seen as providing very detailed points in the problem space, but the relationship between those points may be unclear. Structured interviews are, at first look, little more than spoken questionnaires, following a list of pre-set questions. As such, they offer very similar advantages and disadvantages. However, as for unstructured interviews, interviewers may be given discretion to pursue interesting answers for further details and also can assist respondents if they do not understand a question, thus offering slightly more flexibility than questionnaires. Semi-structured interviews are a compromise approach between structured and unstructured interviews and attempt to offer the most flexibility, while also maximizing the quantifiable data that can be compared across respondents.

Whichever type of interview format is chosen, they are generally time-intensive, requiring one-on-one time between the interviewer and the interviewees, so are only really practical for small samples. They also have to be conducted careful to ensure that the interviewer does not prejudice the data gathered by allowing his or her own opinions to influence or lead the respondents.

User trials, most often in the form of user observation sessions, typically involve watching the users perform the task of interest, either using an existing product or a prototype. As for interviews, they are also time-intensive, and often require the use of specialist equipment for recording and analyzing the sessions. The observers also have to be aware of the need not to interfere in how the user performs the task, to avoid influencing the data collected.

Focus groups are a current favorite among market researchers. They offer feedback from many users in a short space of time, and so are considered good value for money. However, the principal weakness of focus groups is that they can be hijacked by a small, vociferous minority who impress their opinions on the other participants.

Ethnographic methods rely on providing the users with recording media, such as cameras, diaries, and tape recorders. The users then use the media to keep a record of what they consider to be important over a period of between a few days to a week. The strength of this approach is that the user is left at complete liberty over what to record, thus preventing the researcher from influencing the outcome. However, the weaknesses are that the user may not record anything that is relevant and also that the data collected needs to be interpreted and this is, in turn, subject to the interpretation of the researcher performing the analysis.

All of the above methods are discussed in more detail in many usability textbooks (e.g., Mayhew, 1999; Nielsen and Mack, 1994) and those books should be read for further details about those methods.

Packaging the Data

Carefully chosen, in-depth information about a limited range of users is arguably the most useful way of providing data to designers and there are a number of possible methods of packaging the user information for designers. For example, short videos of target users—perhaps depicting their lifestyles, or using or talking about particular products—provide designers with greater insights into the needs and aspirations of users. Such dynamic illustrations can be effective in inspiring designers to formulate inclusive solutions.

Multimedia snapshots supplement imagery—illustrations, photographs, and videos relating to targeted users, their needs, aspirations, and use of products—with short textual descriptions and other complementary information. This is often presented in the form of text-based stories, scenarios, or storyboards representing different users interacting with a particular product or service. Such accounts offer immediate means of assessing a variety of ways and situations in which a product will be used or accessed. It can be a powerful technique if care is taken when building up user profiles based on actual user data or amalgams of individual users, constructed to represent the full range of target users and contexts of use.

Anthropometric or accident data can be encapsulated using graphs, charts, and tables. These can be effective for revealing trends and relative values, and for communicating more vividly what may be perceived as overly "dry" information in designer-friendly formats. Reference tables and other conventional data formats are particularly suited to showing absolute values, but these may not inspire designers or be readily understandable to them. These are all examples of methods of packaging the information for general product design purposes.

ESTABLISHING A BASELINE FOR THE DESIGN

It is almost impossible for a designer to begin designing accessible products without some degree of preparation. The most logical place to start is establishing a baseline from which to begin designing the new, more accessible product. If re-designing an existing product, the baseline is the accessibility of the original product. If designing a new product for which competitors exist, the baseline is the accessibility of those competing products. Finally, if designing a brand new product, the baseline is harder to establish, but will most probably be the accessibility of the nearest kind of product, whether in-house or from a competitor. The question then is what should this "baseline" actually be?

The answer to this question is quite complex, but getting to the answer is surprisingly straightforward. It begins with identifying each of the features and functions of the product. So, for a washing machine, for example, the functions could include:

- load and unload the washing;
- put in bleach, washing powder (or liquid), and fabric conditioner;
- select the water temperature (hot, warm, cold, etc.);
- select the program (cottons, woolens, delicates, etc.); and,
- select any special functions (beeper volume, quick dry, etc.).

Each of these functions is associated with specific features on the product, each of which has its own set of attributes (shown in parentheses):

- load and unload the washing—door (size, weight, orientation, stiffness of latch, finger grip shape and size), end-of-washing-cycle alert (beeper);
- put in bleach, etc.—dispensing tray (size, weight, orientation, finger grip shape and size, labeling of compartments for each liquid);
- select temperature—dial (size, stiffness, shape, labeling);
- select program—dial (size, stiffness, shape, labeling); and,
- select any special functions—buttons (size, labeling, LED).

Each of the attributes of the product features affects its overall accessibility. For example, a door that is too small will force users to load items one at a time, involving many repetitive bending movements. A door that is too big may be too heavy to move or have a very stiff latch to keep it from opening during the wash cycle and so on. The next step is then to gauge the severity rating of the difficulty that each feature represents. This rating will be a weighted sum of:

- the frequency with which the difficulty is encountered (often encountered down to rarely);
- the level of effort required to overcome the difficulty (great effort through to little or no effort);
- the (lack of) presence of alternative methods of accomplishing the task (lots of alternatives through to very few or no alternatives); and,
- the effect of not being able to use the feature (product becomes unusable down to use of product essentially unaffected).

The precise weightings attached to each of these variables will depend on the nature of the product being evaluated. Finally, a priority rating is then attached to each difficulty. This is another weighted sum, this time of:

- the difficulty severity rating (extreme difficulty down to little or no difficulty);
- the number of potential customers affected (lots of people down to very few);
- the ease of fixing the difficulty (easy through to very difficult); and,
- the cost of fixing the problem (cheap through to very expensive).

At this end of this process, the designer should be left with a list of prioritized features to re-design. In the process of re-designing the product, the designer will be faced with a number of possible options for each feature:

- **Can this feature be removed?** If the feature is not essential to the product and seldom used by any of the users, it could possibly be removed.
- **Can this feature be changed to make it more accessible?** For example, if a push-button control is too small for some users to press, the designer should consider making it larger.
- **Can a complementary method of offering the functionality be added?** For example, if users are having difficulty telling that a washing cycle has finished because the beeper is difficult to hear, then maybe a countdown timer or a flashing LED display could be added to supplement the beep.
- **Can the functionality be offered in an alternative way?** For example, if the program select on a washing machine is currently a dial, but too many users find the dial difficult to operate, then perhaps it could be replaced by a push-button control.
- **Can an auxiliary aid (or assistive technology) be offered to supplement to feature?** This approach should really be the option of last resort as the solutions are often comparatively expensive for the users affected and also comparatively unusable. In the case of a washing machine, one option may be a speech activated control system for someone experiencing significant difficulty operating the controls, but such a system would probably double or triple the cost of the machine.

Wherever possible, each of the decisions made by the designer should be double-checked to ensure that the changes to the product do actually increase the accessibility of it and do not introduce some new and, most likely, unexpected difficulties.

IDENTIFYING DIFFICULTIES

As discussed above, central to designing for accessibility is the notion of identifying difficulties that users may experience when using a product. There are several possible techniques for achieving this. Note that although the term "designer" is used throughout this section, the methods described may also be used by usability or accessibility practitioners performing the product accessibility evaluations.

Self-Assessment Method

The fastest method for a designer is self-assessment, in other words the designer tries to visualize a user and his or her capabilities, and then attempts to estimate whether the design is accessible for that user. The advantage of this approach is that it is very fast and cheap to implement—it just requires the designer to stop and think for a few moments. In many circumstances within companies this is the default method of assessment.

The disadvantages, though, should be obvious. Unless the designer has extensive experience of users with different levels of capability, whether from the aging process, specific medical conditions, or physical trauma (e.g., accidents, stroke, etc.), then the designer's estimates are likely to be wildly inaccurate and little more than guesses. In the extremely unlikely event of the designer having such experience, the designer is still probably not able to imagine all possible ways of using the product, let alone assess each of them. What is more, two different designers working on the same design team are highly likely to come up with completely different assessments of the level of accessibility of the same product feature.

Structured Self-Assessment Method

It is possible to add structure to this analysis to try to make it more repeatable and consistent. For example, the assessment can be structured along the lines of a traditional task analysis. Basically, the designer decides on a task that is to be accomplished by the user. This task should ideally be one of those foreseen when the design specification was being formulated, in other words one of the tasks for which the product is being expressly designed. The designer then establishes a scenario of use, that is the environment in which the product is being used, specifically any items that affect the use of the product but are not intrinsic to the product. Returning to the example of the washing machine, the scenario of use would define the layout of the kitchen, the location of the washing powder, fabric conditioner, and so on.

Within that scenario of use for the defined target task, the designer would then break down the interaction into the most basic component steps. For example, putting fabric conditioner in the washing machine dispensing tray would probably involve the following steps:

1. locate fabric conditioner bottle;
2. pick up bottle;

3. unscrew lid;
4. grab dispensing tray handle;
5. pull tray open;
6. release tray;
7. tip up bottle;
8. monitor level of conditioner;
9. tilt bottle back to vertical;
10. push tray closed; and so on

For each of these steps the designer can then identify the parts of product features involved and thus the functional capabilities required of the user to be able to interact with those parts. From this, the designer should be able to derive a list of functional capabilities required and, consequently, a good idea of what kinds of capability losses or impairments would affect the ability of the user to interact with the product to complete the task. We shall return to this basic approach later in this chapter.

A significant amount of up-front effort is required to break the interaction down into such small steps. However, the basic task analysis structure (i.e., identifying the component steps involved in interacting with the product) is a common design practice and may well already feature as part of the design process used to develop the product. This may or may not include the user functional capabilities required to perform each component step. If it does not, then it is a comparatively straightforward modification to add some consideration of the user capabilities required for each component step of the task.

Putting such structure on the self-assessment method does improve the reliability and repeatability of a designer's estimate of the accessibility of a product. However, research has shown that this improvement is still far from the level that would normally be considered satisfactory. This is especially true when considering the results from assessments performed on the same product by different designers. Beyond adding the task analysis structure to the assessment, there are other options for improving the reliability of the product accessibility evaluations, for example the use of user models.

User Models

Another approach to improving the reliability of the assessment is to provide the designer with detailed data or models of the user that provides a breakdown of user capabilities by percentage of the population. Therefore, if the designer knew that a force of 15 Newtons was required to open a washing machine door and 85 percent of the population was capable of exerting that force through their arms, then clearly 15 percent of the population would experience difficulty opening that particular washing machine door. If the same data showed that reducing the force to 12 Newtons presented difficulties to only 7 percent of the population and also did not compromise the safety of the door (it would not do for it to come open suddenly in the middle of a spin cycle), then the designer would have established a specific design target to aim for.

The disadvantage of this approach is that the amount of data that would need to be made available to the designer would be so immense that there would be a real danger of the designer not being able to know where to start for fear of drowning in numbers. Only if accurate biometric computer models of all of the stages of interaction could be established (itself no mean feat) and then correlated with a smart database design to automatically find the relevant population data, could such an approach be considered realistic for all aspects of the product's use.

The other risk is that the designer may get so caught up in the details of the assessment that he or she ends up focusing on trying to tweak the data to get the best possible result at the expense of not taking a step back, looking at the overall picture and trying more radical solutions. The net effect would be a massive effort to establish a local optimum and missing the global optimum. This is otherwise known as trying to make the best of a bad solution instead of trying to find the best solution.

Please note, though, that this approach does have a good track record when considering functionally simple products, such as designing cutlery or door handles. In these circumstances, the interaction with the product is straightforward to understand and consequently lends itself to simple, very focused data about the user population. For example, for both of these products the designer would be most interested in hand sizes and finger and arm strength. In some cases, the data could even be developed into simple checklists that outline the physical properties required to ensure that the product is accessible. Example checklist items could be:

- Are all font sizes at least 16 point?
- Are all finger clearances at least 18mm in size?
- Is the total weight of the product less than 1.2 kg?

However, the data requirements increase exponentially as the complexity of the product increases and these approaches would be unlikely to work when considering more high-tech products. The other significant issue is that the numerical values used in such checklists are invariably derived from anthropometric and ergonomics data tables. As such, they usually represent a percentile proportion of the population, such as the 98th percentile, for example. By default, such values will automatically exclude 2 percent of the population (in the 98th percentile case). There is no easy remedy to overcome this limitation as data is simply not collected in absolute terms of the whole population, e.g., the absolute maximum thumb size that could possibly be encountered.

Simulation

The other approach to helping designers perform assessments themselves is simulation. The nature of the simulation aids depends on the type of product being considered. For example, for tangible products (hardware products, if you like), simulation typically involves the designer wearing simulation aids that emulate physical impairments and thus reduce the designer's innate physical capability. The simulation aids can be varied according to the user capabilities being investigated. Table 8–1 provides

TABLE 8–1

Examples of Simulation Aids and the Impairments They Simulate

Vision Capability

Aid:	Simulates:
Goggles smeared with grease	▪ Cataracts
Spectacles with incorrect strength for the user	▪ Low vision
Blacked out lens	▪ Loss of vision or blindness
Lens with spots painted on	▪ Age-related macular degeneration (a common cause of vision loss in older adults)
Colored lens	▪ Color insensitivity or blindness

Hearing Capability

Aid:	Simulates:
Ear mufflers or defenders	▪ Loss of hearing
Headphones playing high-pitched bells	▪ Tinnitus

Motor Capability

Aid:	Simulates:
Gloves	▪ Loss of tactile feeling in fingers, impaired fine motor control
Wrist weights	▪ Reduced arm strength
Joint supports (wrist/knee/elbow, etc.)	▪ Restricted joint motion

examples of different possible aids that are readily available to designers to acquire or make.

Note the absence of any kind of cognitive impairment simulation aid. Cognitive impairments are very difficult to simulate reliably. There are options available, such as distraction techniques or stress techniques, which both basically involve trying to either distract the designer from the task in hand through imposing additional calls on the designer's attention. A common technique would be to play a soundtrack of an argument or even white noise. The problem with these options is that they can quickly become stressful for the person performing the assessment and are, therefore, generally best avoided.

For software products, it is possible to offer another type of simulation aid. While the aids described above can still be used, it is possible to complement them with software models of other types of impairment. For example, consider a typical mouse and keyboard input set-up for interacting with a graphical user interface program. Spasms or tremor could be simulated by writing a program that adds those behaviors artificially to the mouse input stream, without the need to build an expensive moving rig to physically displace the mouse. Unfortunately, though, such simulators are not currently available outside of research labs.

Simulation is a very effective method for identifying potential accessibility difficulties in products. It is also generally very repeatable and consistent across different assessors, with the possible exception of two assessors with grossly different physical characteristics. For example, if one assessor is significantly stronger than another one, then the aids may need to be tailored to compensate for those differences, e.g., by adding more weight or increasing their stiffness and so on. Simulation is also generally very quick and makes the designer think about how a product is really going to be used.

The disadvantage with simulation is that it still does not offer a genuine insight into how a real user would view the product. It only offers an insight into the consequences of the physical impairment and not how the product relates to what the users really want or how the product would impact upon their lifestyle. For that level of additional insight, there is no option other than involving users.

User Trials

User trials are generally regarded as the gold standard for evaluating the accessibility of a product. However, they are not without their drawbacks—principally the difficulties often faced by companies trying to recruit users and the cost and time involved in setting up and running the user observation sessions. These issues are worth investigating in depth and will be discussed in detail in later chapters on "Involving users in the design process" and "Conducting sessions with users" (Chapters 10 and 11 respectively).

SUMMARY

Designing for accessibility is, by definition, a strongly user-centered design activity. As such, it is important that designers be provided with access to users or else to accurate and reliable data about them. That aside, the basic design practices of the designer most likely do not need to change substantially. It is most likely that they just need to be expanded slightly to allow for the new design requirements that arise from the desire to design an accessible product. Perhaps the most significant addition to traditional design models is the need to consider the effect of all design changes on the overall accessibility of the product.

KEY POINTS

- It is imperative that the user wants and needs for the product are identified accurately.
- Designing for accessibility relies on the ability to identify potential accessibility difficulties with a product.
- Those difficulties need to be prioritized and then fixed or removed.

9

Case Study—Designing for Accessibility in Practice

This chapter presents a brief case study of the re-design of an interface for a quite complex prototype rehabilitation robot. Rehabilitation robots were a popular topic for research in the 1980s and 90s. It was commonly thought that if robots could be designed that could perform routine everyday tasks, then they could be used to assist someone with a severe motor impairment and who consequently experienced difficulty performing many of the tasks that are required in everyday life, including for employment.

However, many of the robots never made it out of the research laboratories. Of those that did, they typically achieved fairly miserable success in the marketplace (Buhler, 1998). The reason for this was that the robots were primarily being designed by mechanical engineers, with limited input from human factors or usability professionals. Thus the principal focus of the research and development activities was dominated by the logistics of the mechanical system development, such as motor power, reliability, materials selection, and so on. The logic behind this emphasis was that building the robots was a stiff technological challenge and so that was where the majority of effort should be focused.

Typically it was only at the last minute that the development teams considered what the interface should be. This was a recipe for disaster. With a couple of notable exceptions, the interfaces were uniformly dreadful. Users who, because of their functional impairments, needed interfaces that were very straightforward to use, found themselves confronted with multiple controls that were poorly explained and often difficult to operate. In this chapter we will examine one such robot and how, for a fraction of the cost of a rival mechanical overhaul proposal, an accessibility study led to the development of a successful and popular user interface design.

BACKGROUND

The aim of the Interactive Robotic Visual Inspection System (or IRVIS for short) was to enable the remote inspection of hybrid microcircuits. Hybrid microcircuits are typically found inside mobile telephones and other handheld devices. They often couple traditional computer-type digital circuits with other non-typical equipment such as antennae, microphones, and the like. The microcircuits are also quite small, often less than an inch square in size, typically with several hundred components on them—see Figure 9–1 for an illustration of an example microcircuit and also an example flaw, in this case a broken wire bond.

At the time that the robot was built, the inspection task was performed by able-bodied inspectors, who handled the circuits under an optical microscope. The IRVIS system was developed because the inspection process is fundamentally a visual task and would-be inspectors were being excluded from potential employment because of the reliance on the manipulating the circuit by hand.

The entire purpose of the robot was to be able to move a camera over the circuits without touching them and thus avoiding the risk of damaging them while they were mounted in the tray used during their manufacture and display a significantly magnified image on a computer screen. This process would replace the manual approach, which involved removing the circuits from the tray and moving them by hand under a microscope. This involved extensive handling of the circuits and, thus, far greater risk of incurring damage to them during the inspections. Note that during manufacture, each circuit would typically be inspected approximately 50 times. The use of IRVIS in the workplace would remove an unnecessary barrier to potential operators with

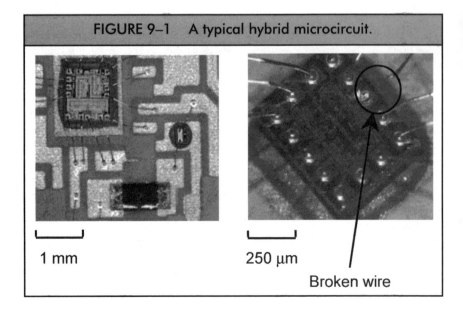

FIGURE 9–1   A typical hybrid microcircuit.

1 mm

250 µm

Broken wire

motor impairments, who do not have the manual dexterity to hold and move the delicate circuitry, but do have the required level of eyesight to identify any defects.

The original interface design for IRVIS was the result of a logical engineer's approach to how to control the robot. It allowed each of the robot motors to be driven to a high degree of accuracy. However, it failed completely to recognize that users did not think about the task in terms of which motor to actuate. Instead, all they cared about was whether they could rotate about a fixed point, or move it *there*.

THE IRVIS PROTOTYPE

The prototype IRVIS system was developed by a researcher at the University of Cambridge (Mahoney, Jackson, and Dargie, 1992). It consisted of a movable tray with four degrees-of-freedom, that is to say it could move forwards, sideways, upwards (for focusing the camera), and rotate in either direction. A digital video camera was mounted on a tilting arm above, with freedom to move sideways also, adding a fifth degree-of-freedom. Figure 9–2 shows the prototype. Combining these five degrees-of-freedom meant that the camera could be pointed at, and focus on, any part of the circuitry from any possible angle.

Each of these five degrees-of-freedom was controlled by a single motor—a perfectly logical solution when trying to provide the necessary utility in the robot. However, this arrangement of five motors, whilst offering all the requisite functionality, resulted in complex kinematics to perform basic inspection tasks. For example, examining whether a wire was bonded correctly on the board involves looking at the bond

FIGURE 9–2   The prototype IRVIS.

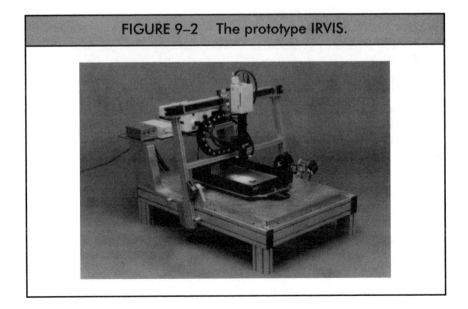

from all possible angles. This in turn involves tray and camera translation, tray rotation and arm tilting—quite a complex combination of movements to co-ordinate. Consequently, a routine inspection procedure could easily involve controlling all five motors simultaneously.

THE ORIGINAL INTERFACE DESIGN

The original interface design followed the same logical patterns of thought as the layout of the robot. The interface was driven by menus. The user had to select the menu that applied to the particular motor to be controlled and then specify how far to move that motor. Again, all of this was perfectly logical—and completely unusable.

User trials at a hybrid microcircuit manufacturer demonstrated the feasibility of the system, in other words the interface did indeed control the motors as intended, but highlighted the all-too-obvious shortfall in overall usability. Put simply, despite offering all the necessary utility, the system was not meeting the needs of the inspectors and a new interface was clearly required. Thus it was decided to re-design the interface through the application of the 7-level model described in the previous chapter.

DEFINING THE PROBLEM

The first stage of applying the 7-level model involves identifying and defining the problem to be solved. From the results of this stage, the design specification is created complete with both social and practical acceptability requirements.

Level 1—Problem Requirements

In this case, what was required was an interface that allowed a wide range of users to control the automated movement of a camera to within a predetermined degree of accuracy, repeatability, and time constraints. The solution also had to meet targets of social acceptability—the users had to feel comfortable using the robot. The specified target users were the existing, nominally able-bodied, inspectors and also users with motor impairments. There were other constraints reflecting the size of the robot, total cost, and development schedules and so on, but the interface requirement was the one that we are concerned with here.

Level 2—Functional Requirements

The functional requirements for the re-design centered on providing the user with the capability to move a camera above and around a microcircuit in such a way that faults in the circuit could be seen and identified. Given that there was an existing interface for the robot and that it appeared to be less than successful, the obvious starting point was looking at what its problems were. The problems with the original interface were immediately apparent. They were principally due to the users being unable to understand and predict the effects of commands entered through the interface and the resulting motion of the robot. The commands were too abstract and distant from the immediacy of manual circuit manipulation, resulting in a lack of feeling "in control." The IRVIS system required a structure enabling intuitive direct control, rather than the more detached supervisory control offered by the original interface.

The next move was straight out of the usability textbooks—find out what inspectors actually did when examining the circuits. To that end, a significant amount of time was spent interviewing inspectors and video-recording them performing routine inspections. The typical inspection procedure involved the inspector moving the circuits under a bright light and watching this through a binocular microscope. Their aim was to see if there were any changes in the reflection of the bright lights across the surfaces of the smooth components. Such changes would indicate the presence of a scratch and the circuit would be scrapped.

On analyzing the data collected, a list of the typical movements of the circuits was compiled. It was quickly found that inspection using an optical microscope involved a number of distinct actions, which may be grouped into five generic categories:

- translation—moving the circuit sideways, or forwards and backwards;
- rotation—turning it left or right, often about a particular point, i.e., the center of rotation;
- tilting—rolling it to look at it from front edge-on to back edge-on;
- zooming (the microscope)—to look more closely at a particular feature; and,
- focusing (the microscope).

DEVELOPING A SOLUTION

Each level of the solution development process was accompanied by user trials with users with motor impairments at a local residential care center throughout the level and a final evaluation period before progression to the next level. This approach provided a clear and unambiguous framework with clearly defined goals for system usability and accessibility.

The Role of the Prototype

An integral part of any user-centered design approach is the use of prototypes to embody the system at each stage of development. There are a number of forms that a prototype can take from low fidelity (i.e., more abstract) representations through to high fidelity (i.e., more concrete) working models. In this case, a variable fidelity prototype for use in the IRVIS interface re-design was developed (Dowland, Clarkson, and Cipolla, 1998). This prototype was in essence a software simulation of the complete system that encompassed both the appearance and functionality of the user interface and the mechanical properties of the robotic hardware—see Figure 9–3.

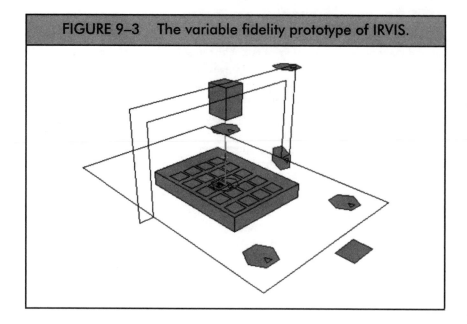

FIGURE 9–3   The variable fidelity prototype of IRVIS.

Users

A total of six users volunteered to participate in the user trials for the re-design of the IRVIS interface. Each participant was a full-time resident of a local care center. The care center was run by a charitable organization that actively sought to find employment for its residents at local companies.

- User 1 was a forty year-old male with muscular dystrophy, a muscle-deteriorating condition that left him with fine motor control, i.e., the ability to complete small, detailed tasks, but limited strength. His preferred computer input device was a trackball. He was non-ambulatory and used an electric wheelchair.

Four users had cerebral palsy. Of those users:

- User 2 was a young male in his twenties who exhibited spasms in his arms, especially when concentrating on a fine task. He was also non-ambulatory and used a manual wheelchair.
- User 3 was a deaf, non-speaking male in his forties. He exhibited generally limited co-ordination, but had good strength. He was ambulatory, using two walking sticks for short distances and an electric scooter for longer treks.
- User 4 was a male in his thirties who also exhibited limited co-ordination, but was ambulatory using a pair of walking sticks for short distances and drove a specially adapted car for longer journeys.

- User 5 was another young male in his twenties. He was formerly ambulatory using walking sticks, but a number of stumbles and accidents had convinced him to use a manual wheelchair instead. He exhibited spasms in his arms and also could not extend his fingers, using the mouse with his hand resting as a clenched fist on top of it. He pressed the mouse buttons by rolling his knuckles onto them.

- User 6, another male in his forties, had Friedrich's Ataxia, a condition causing constant body tremor. Like User 5, he could not extend his fingers and used the base of his palm to move the mouse. He also rolled his fingers forward to operate the mouse buttons with his knuckles.

All six users had either 20/20 uncorrected visual acuity, or wore corrective prescription spectacles and thus were the type of users for whom IRVIS was intended.

Level 3—Output to User

After developing a basic model of the system, work focused on the problem of defining a minimal, but sufficient, representation of the system for the user to be able to interact with. This version of the revised interface showed an overview of the robot and a simulated camera view of the tray of microcircuits to be inspected—see Figure 9–4. The user was able to select control over any one of the robot's individual motors and to drive them by clicking and dragging the mouse cursor on the hexagons that represented the motors.

The users were asked to predict the machine's behavior as a result of their input. Initially, they had some difficulty understanding what was being presented to them and it quickly became clear that apparently simple details can make a sub-

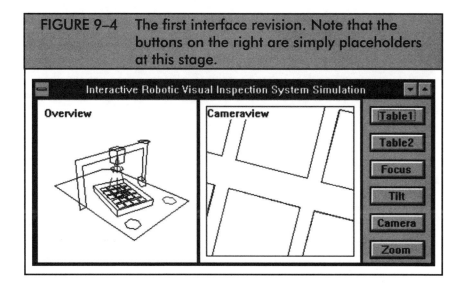

FIGURE 9–4   The first interface revision. Note that the buttons on the right are simply placeholders at this stage.

stantial difference to the overall usability. Small changes had significant effects of the overall understandability of the interface. For example, the addition of a view cone (literally a cone shaped animation emanating from the camera down onto the circuit tray) in the left-hand view (the robot overview) showing where the camera was pointing helped the users gain a better understanding of the orientation and positioning of the right-hand image (the simulated camera view). Other subtle modifications, such as changing the color of the motor that had been selected from black to red also helped.

Specific modifications based on the motor impairments of this particular user group included increased tolerance for the selection of the specific motors. This was achieved by allowing a click just outside each motor to be treated as if it was a click on the motor. In addition, the mapping of mouse movement to motor rotation was adjusted to filter out any effects of tremor or spasm. This was the result of several iterations of different ratios of movement to rotation. Once all of the users were able to complete this task without assistance, the design moved to the next level of the 7-level model.

Level 4—User Mental Model

Having established a representation that afforded sufficient feedback to the user, the next step was to add the full functionality of the IRVIS robot to the software simulation, while ensuring that the interface retained its overall understandability. The user trials for this stage of the re-design were to ensure that the simulated robot response to user input was consistent with that of the actual hardware. The robot was connected to the computer and the users were initially asked to repeat the same procedure as for Level 3, only this time predicting what the robot would do in response to their actions. Once the users were comfortable controlling the robot, new functionality was added to the interface that replicated the five basic actions that had been seen from the inspectors: translation, rotation, and so on. Only when all of the users were able to implement each of these actions did the design progress to the next level of the re-design process.

An interesting side-note came out of this stage of the work. Through the new interface offering much more direct feedback to the user, it became clear that poor mathematical assumptions had been made in the original interface about how the camera rotated about a particular point on the microcircuits. Initially it had been thought that the difficulty in rotating about a point was because of mechanical inaccuracies in the robot and an expensive mechanical re-design had been proposed by another set of consultants. However, on further investigation, it was found that no allowance had been made for the variation in height for different components on the circuits. The underlying control system for the robot and the simulated prototype were adjusted accordingly and the expense of the (unnecessary) mechanical re-design avoided. All of this was achieved simply because the users were able to anticipate what should happen in the interface and what they saw happening instead.

Level 5—Input from User

The final stage of the re-design concentrated on assessing the ease of interaction between the user and the robot, identifying particular aspects of the interface that required modification. The task in the user trials changed from "What will the robot do now?" to "Can you accomplish this goal?"

The "goal" was the inspection of actual hybrid microcircuits and specifically each of the types of actions performed by the inspectors, i.e., translation, rotation, tilt, zoom, and focus. From each of the previous stages, it was clear that all of the users wished to interact as directly as possible with the circuit and not with the motors. Consequently, the individual motor controls were replaced with buttons offering each of the five generic inspector movement types—see Figure 9–5.

The size and direction of each of these inputs was directly proportional to the magnitude and direction of the mouse movement. Thus the user could manipulate the circuit directly and the interface became easier to use. The speed-of-response parameters were also adjusted until the users were comfortable with the "feel" of the robot. The interface supported the setting of velocity and acceleration profiles, for each individual user.

VALIDATING THE SOLUTION

While each of the levels involved in the development of the new interface design involved user trials, it was still necessary to perform a final confirmatory set of trials. These trials involved the users performing a representative series of inspection tasks derived from the observational study of inspectors performed as part of Level 2.

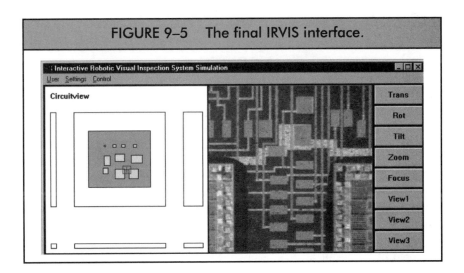

FIGURE 9–5   The final IRVIS interface.

Level 6—Functional Verification

All of the users were able to navigate around the circuit tray without difficulty and within the time limit allowed. Likewise, all of the users were able to perform all of the other tasks. These included:

- Locating specified components on specified circuits—each typical tray of circuits contains 20 individual circuits, all of which are identical, so the users had to be able to identify which circuits they were looking at and then which particular components to find the specified ones;
- Finding scratches on the surfaces of circuit components—involved tilting the camera over the circuit and looking for differences in the reflected light from the surface;
- Inspecting wire bonds—wire bonds typically arc up from the circuit board to the top of the components, so are three-dimensional structures in effect.

Inspecting a bond involves moving the camera along its length and rotating around it to look for any breaks in the bond, which can often only be seen from one particular angle at one particular location. This is a very complex task to perform and all of the users were able to accomplish this task, along with all of the other tasks.

Level 7—Solution Validation

Qualitative feedback from all the users was extremely favorable. Each user found the new interface easy and intuitive to use and all completed the tasks with a minimum of guidance. No user complained of the speed of response of IRVIS being too slow. This was an important result, because it had been previously thought that IRVIS was mechanically under-specified. A simple analysis showed why this was so. The original interface only allowed the use of one motor at a time. The new interface allowed potentially all five motors to be used simultaneously. The increased power available to the user significantly improved the overall speed of response.

IN CONCLUSION

The IRVIS example is instructive because it could just as easily be described as a *usability* study as it can an *accessibility* one. Many of the issues that arose from considering design for users with motor impairments were equally applicable to users without such impairments. The concept of direct control, for example, and the resultant increase in robot performance from coordinating the use of the motors simultaneously are results that any good usability study would have found. The only specific differences that were seen were the slightly more generous tolerances for the selection of the individual motors in the early stages of the re-design and also the specific acceleration and velocity profiles developed for each user in the final set of trials. Note, though, that while this accessibility study (performed with motor-impaired users) has highlighted many usability problems, it is extremely unlikely that a usability study (performed with able-bodied users) would have highlighted the accessibility ones.

The purpose of including this case study was to show how little the user-centered design process had to be adjusted to include users with motor impairments. It also demonstrates that with appropriate design and technology, users with severe motor impairments could perform an extremely delicate motor-based task. This reinforces the notion that a person is not fundamentally disabled by his or her own impairments, but is disabled by inadequate consideration and support for someone with their particular combination of functional capabilities. In other words, nobody is innately *disabled* (see Figure 2–1 for a reminder of the World Health Organization's interpretation of disability). Certain people may have reduced capabilities (*impairments*), but these only become *disabilities* if the products required to complete a particular task do not support those levels of capability. This may seem like a purely semantic difference on the surface, but is a very important distinction. It reinforces the notion that design for accessibility is not simply "design for the disabled," because disability is principally a result of inadequate accommodation. Design for accessibility is about ensuring that people do not become "disabled by design."

Looking back at the IRVIS interface design, it is worth noting how each stage of the development process was accompanied by user trials. While user trials are generally considered to be expensive and time-consuming, in this case they proved to be extremely cost-effective. By developing an interface that allowed the users to operate the robot to its full potential, a costly re-build of the mechanical prototype was avoided. The costs associated with that re-build were estimated to outweigh the costs of the user trials by an order of magnitude. In other words, the user trials cost less than 10 percent of the re-build. To say that this was a significant saving would be something of an understatement, especially when you consider that a re-built robot would still need a new interface developing for it.

KEY POINTS

- This accessibility user study highlighted many usability difficulties. It is unlikely that the reverse would hold true.
- The use of user trials avoided a costly re-build of the robot. The cost savings were very significant.
- A new employment opportunity was created for people with motor impairments. The cost of developing an inclusive, accessible interface was no more than developing a usable one.

10

Involving Users in the Design Process

Back in the opening chapter of this book, the assertion was made that user-centered design was arguably a company's best bet for designing a successful product. Central to the concept of user-centered design is, of course, working with users. Ideally, users should be involved actively throughout the design and development process, whether taking part in user trials, as full members of the design team, or simply chipping in with ideas and opinions. This is the "gold standard" approach. However, not all companies have the resources to support such close and continuing contact with users over what could be a considerable period of time. Under these circumstances, companies are much more likely to focus on critical "gateway" user trials, i.e., trials that determine whether the product has met its objectives and can proceed, or is deficient somehow and needs modification.

At the very least, users should be involved right at the outset of the design process. They should be actively involved in defining the very core of the design specification. Similarly, they should also be involved in the final phasing of testing to verify that the final product is accessible and usable. This is the very minimum level of user involvement that a company should consider. In practice, though, involving users at the extreme ends of the design process leaves a potentially alarming chasm in the middle, where most of the decisions that shape the product and its functionality are being made. This yawning chasm can be bridged to some degree by making use of additional information resources on accessibility (such as tools, guidelines, and standards) and domain experts, but these are primarily substitutes for the real thing, the users. Therefore, wherever possible, user trials should be held at all key points in the design process. Deciding when to involve users in the design process is only the start. Deciding who to involve, and how, are the next issues to address.

WHICH USERS SHOULD BE INCLUDED?

The types of users to include in trial sessions are dependent on the product being evaluated and the known target audience. For instance, if considering a product with no auditory component in its use, such as cutlery for example, it may not be necessary to include users who are deaf or hard of hearing. The ideal situation would be to test a product with as many users as possible and as wide a range of impairments and capabilities as possible. In practice, however, time and funding constraints may well dictate that the fewest possible users are recruited to participate in the trials and the trials last the least possible time.

Of course, if too few users are recruited and the sessions are too short, then it is highly likely that accessibility problems will go undiscovered. If just one of those is a critical flaw, i.e., makes the product unusable for a significant number of users, then the product will fail. What is required, therefore, is to find users that offer the best "value for money" in the user trial sessions. The users that are of most interest are those who should (or need to) be able to use the product, but who might not be able to and thus will be most affected if the product is not accessible. One of the best approaches to adopt is a 3-stage process.

- Stage 1—Identify the facets and characteristics of the users that are likely to be of most interest;
- Stage 2—Decide on the user sampling method;
- Stage 3—Find and recruit users.

Stage 1—Identifying the User Characteristics of Interest

The easiest way of identifying which users may be affected is to break down the interaction into its common component steps and think about what the user has to be able to do to complete each step, e.g.:

Software scenario example:

A user is using a word processor and decides to print the document being worked on. The "Print" dialogue box has popped up on the screen and the user needs to select the "OK" button.

Steps:
1. Recognize that a dialogue box has appeared—either see it (vision) or hear the notification sound (hearing)
2. Find the available buttons to choose between (vision)
3. Distinguish between the buttons to choose correct one (vision, cognitive/learning)
4. Decide which one to activate (cognitive/learning)
5. Activate it (vision, motor)

CHAPTER TEN

Hardware scenario example:

A user is trying to withdraw cash from an ATM. The ATM is prompting the user to answer the questions "Which account do you wish to withdraw your cash from?" The user has to press the relevant (hardware) button to proceed.

Steps:
1. Read the prompt—(vision, cognitive/learning)
2. Find the available buttons to choose between (vision, maybe motor)
3. Distinguish between the buttons to choose correct one (vision, cognitive/learning)
4. Decide which one to activate (cognitive/learning)
5. Activate it (vision, motor)

The similarities between software and hardware activation of a button are obvious. Note that each of the principal impairment classifications (sensory, motor, and cognitive) could potentially encounter a "showstopper" accessibility difficulty in these scenarios. Also note that the use of assistive technology requires these steps to be gone through again and amended accordingly. For example, the use of screen-reader software in the software scenario removes the vision demands on the user, but replaces them with hearing ones.

Clearly considering the needs of just one user group will only identify the issues for other users of that same group. If resources are only available to work with one or two users, one option is to find users with a range of impairments. Older adults, for example, may have limited dexterity (motor impairment) through arthritis, as well as low vision from age-related macular degeneration and slight hearing loss. It should also be borne in mind that some users may require highly specialized or customized solutions to meet their needs. Such solutions may not necessarily be applicable or beneficial to users with lesser impairments. It is preferred practice to find users who should be able to use the product but are experiencing, or are likely to experience, unnecessary difficulties in doing so.

Note that while the above discussion is focused on capabilities, it is eminently plausible, and indeed reasonable, to extend the user characteristics to include anthropometric details of the user. For example, when considering a product designed for older adults it is worth noting that the average height of someone over 65 is over 2 inches less than the average height of younger adults (Smith, Norris, and Peebles, 2000). If such differences are overlooked when deciding, for example, on the viewing height of a screen, then older users may find themselves unable to use the final product.

Stage 2—Deciding the Sampling Method

Having identified the user characteristics of interest, i.e., the sampling criteria, the next step would appear to be to find appropriate users. However, there is an intermediate step, that of deciding on the sampling method. This is an additional step that arises because of the unique requirements of designing for accessibility. There are many possible approaches for sampling potential users, but for our purposes here, we shall focus on four:

1. sampling users by condition;
2. sampling users by capability;
3. sampling by homogeneous user groups; and,
4. sampling by heterogeneous user groups.

Between them, these four sampling methods can be thought of as a matrix, as shown in Figure 10–1. The choice is not really one from four, but one from the first two and one from the second two options, e.g., a group of similar users with a particular medical condition, or a group of diverse users with a variety of capabilities.

Sampling Users By Condition

The most obvious approach to sampling is to identify users based on their medical condition, such as cerebral palsy, age-related macular degeneration, or Parkinson's disease. The advantage of this approach is that someone's medical condition is a convenient label for identifying potential users. Not only are most users aware of any serious condition, especially one that affects their functional capabilities, but it also makes locating users easier. For example, many charitable organizations are centered on serving the needs of individual, specific medical conditions.

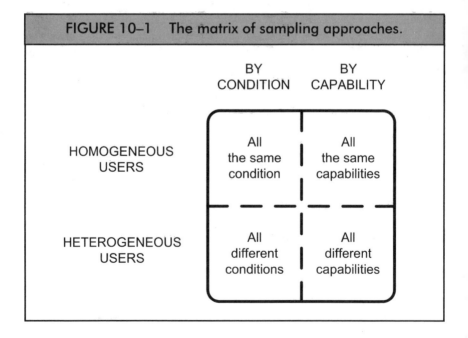

FIGURE 10–1   The matrix of sampling approaches.

	BY CONDITION	BY CAPABILITY
HOMOGENEOUS USERS	All the same condition	All the same capabilities
HETEROGENEOUS USERS	All different conditions	All different capabilities

The principal disadvantage of this approach is that many of these conditions are highly variable in terms of their impact on the user's functional capabilities, and so a degree of user capability profiling may still be required to match users to the desired user characteristics. In addition, designers are not medical doctors. If sampling by medical condition is the preferred approach, then designer-friendly summaries of the effects of each condition have to be provided to the design team.

Sampling Users by Capability

The alternative approach to sampling users is not to focus on their medical condition, but to look instead at their capabilities. The advantage of this approach is that the users are automatically close to the desired user characteristics identified in Stage 1, since those characteristics now form the basis of the sampling procedures. This supports the concept of impairment of function being the most important factor in interaction with a product, rather than a medical condition.

The disadvantage of this approach is that more user capability profiling, e.g., finding out the levels of each of the seven capabilities discussed earlier, is required at the outset to establish where each user sits in the capability continuum in order to match the users to the desired characteristics. This can be a time-consuming process, especially if there are a significant number of users to be sampled.

Homogeneous User Groups

A homogeneous one offers a greater likelihood of generating statistically significant data when compared to a similarly homogeneous control group and the results are thus more likely to find favor within the research and design communities. However, the weakness of this approach is that it needs to be repeated at potentially many different points in the capability continuum to provide a wider view of the general population. There are also practical difficulties in finding the required number of users with very similar profiles to participate in the experimental sessions. Also, for complete homogeneity, additional factors beyond functional capabilities, such as personal experience, knowledge, and motivations need to be considered. Finding such homogeneity is not impossible, for example in a large company where cohorts of employees go through the same training programs, but is nevertheless a significant challenge in most circumstances.

Heterogeneous User Groups

The heterogeneous user approach argues that ideally the users sampled for participation in product research and design should represent the full range of end-user capabilities that can reasonably be expected to be found in the intended target population. However, to achieve statistical significance at all possible levels of capability across the target users would require a large number of participants. For example, if the capabilities involved in an interaction are vision, hearing, and dexterity (as discussed above), and each of these have 10 levels of severity, and there need to be 5 users at least (7 or 8 would be a more typical minimum—see Nielsen, 1993) for each level, then 150 participants are required. If it is considered that there may be interactions between the capabilities and the ability to interact with the product, then this number could increase exponentially as 10×3 levels of severity becomes 10^3 combinations and so potentially 5,000 participants could be required. So, methods of reducing the number of users are needed.

The most popular approaches to sampling issues are to either find users that represent a spread across the target population, or else to find users that sit at the extremes of that population. The advantage of working with users that represent a spread across the population is that they ensure that the assessment takes the broadest range of needs into account. The disadvantage, though, is that there is not much depth of coverage of users who may experience difficulties in accessing the product.

The advantage of working with the extreme users is that the user observation sessions will almost certainly discover difficulties and problems with the interaction. However, the disadvantage is that there is a real danger of discovering that particular users cannot use the product, and little else beyond that. For example, giving an instruction book to a user with complete sight loss yields the obvious difficulty arising from the inability to read the text. However, subsequent questions about the content of the instructions are not possible because of the over-riding difficulty of reading. This is of only limited value in an assessment such as this, as the difficulties encountered by the extreme users are comparatively predictable and provide little information about how many other users may or may not be able to use the product. It could also be argued that such users may reasonably be expected to make use of assistive technology to help access particular products.

Of more use is to identify users who are more likely to be "edge-cases," those who are on the borderline of being able to use the product, and who would commonly be accepted as should be able to use the product (Cooper, 1999). Going back to the example of someone with a visual impairment attempting to read an instruction book, while someone with complete vision loss would certainly not be able to use the instructions, someone with only partial sight loss may be able to do so. Even more interestingly, that person might be able to read some bits and not others and thus it is possible to begin to infer a wide range of very useful data from such a user. Further, if the user cannot read the instructions, then it may be inferred that any user with that level of sight loss or worse, will not be able to use them, automatically encompassing the users with complete sight loss in the assessment of product exclusion.

Figure 10–2 summarizes the different approaches to sampling the users. The implication of this is that whichever group of users participates in the assessment, it is important that their capability profiles are known so that it is known how many users share the same characteristics.

Stage 3—Finding and Recruiting Users

Having decided which types of users should be included in the product assessments, the next stage is to find suitable participants. For traditional usability assessments, the users would typically be customers (existing or projected future customers) or employees, and would often be readily to hand. However, when considering users

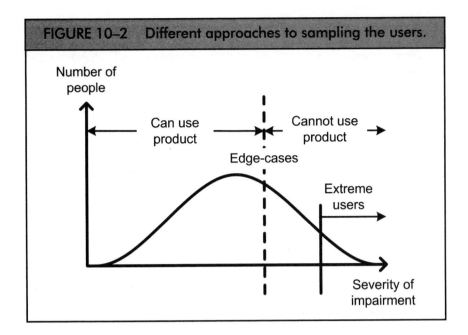

FIGURE 10–2   Different approaches to sampling the users.

Number of people

Can use product ← → Cannot use product →

Edge-cases

Extreme users

Severity of impairment

with a wider range of capabilities, it is often necessary to commit explicit effort and resources to seeking out potential participants, quite possibly from other sources. This is not to say that the usual sources of users should be ignored—far from it. Customers and employees (depending on whether the focus of the design is a new product or improvements to the workplace, respectively) should still be the first resource considered. However, it is unlikely that sufficient users matching the desired user characteristics from Stage 1 will be found, in which case additional pools of potential user trial participants need to be found.

Good sources of users include charities specializing either in helping older adults or people with functional impairments arising from specific medical conditions. Social clubs are often a good source of more active older adults, whereas residential care homes often house potential participants with lower levels of capabilities. If necessary, putting advertisements in local newspapers can also work.

Do not give up. User groups *can* be found in most towns and cities, but effort does need to be expended in trying to find them and then to identify candidate users who match the required user characteristics and sampling profiles. Once a good local source of users has been found, e.g., a day-care center, it is often worthwhile setting up a formal agreement with the center. In this way, the process of finding users for later projects can be streamlined. It also offers an opportunity to the center for publicity activities—showing how it is involved in work to make products more accessible for its members.

Finding users is not the end of the recruitment process, though. For example, if a potential pool of twelve users has been found and there is enough funding (or time) to only include five of those users in a series of trial sessions, then some kind of pre-screening process is required to identify the potentially best participants. An ideal pre-screening exercise is a reduced version of the planned user trial. Thus, if the trial was expected to explore five potential tasks, asking the user to perform one similar (but not identical) task would make a good pre-screening exercise. This mini-trial would help the person doing the screening assess whether the user is suitable based on, for instance, the following criteria:

- Did the user appear to enjoy performing the task?
- Did the user indicate interest and/or willingness in participating in more trials?
- How well did the user follow the instructions?
- Did the user have interesting and useful opinions about the task?
- Did the user communicate those clearly?

Users should not only be selected according to how well they meet the desired user characteristics, but also how enthusiastic and communicative they are. These last two qualities are essential to helping the design team get the best possible feedback from them.

ANALYZING THE RESULTS

At the end of any user study, a wide range of data will usually have been collected that requires careful analysis. Many of the techniques and measures from usability and ergonomics practice can be used, but care needs to be taken in how they are applied. Here we will examine two areas that need very careful consideration: the application of statistical analyses and defining what constitute acceptable results.

The Application of Statistical Analyses

It is important for professionals in the field of designing for accessibility to note that practical limitations, principally involving the variable availability of individual users and the comparatively small sample set that often provides results, can restrict the usefulness of detailed statistical analysis. Traditionally, building extensive knowledge about users frequently requires access to many users, or else a smaller set of users over a longer period of time (longitudinal studies), very often with weekly experimental sessions. Since time and financial constraints, combined with the possible difficulty in finding and recruiting users, tend to eliminate the possibility of working with large sample sizes (i.e., lots of users), longitudinal studies with fewer users is the approach most commonly adopted.

Although we have defined designing for accessibility as not being specifically designing for people with disabilities, in practice many of the users that take part in user trial sessions are likely to be disabled, sometimes severely. Users with severe disabilities may experience changes in their capability over time and also may require medical treatment such as surgery during an experimental series. They may become fatigued easily, despite extremely high motivation, and sometimes cannot complete trials or conditions. Like other voluntary users in long-term studies, they may often choose not to attend experimental sessions.

It is necessary for experimenters to run the trials on a long-term basis and to develop a working relationship with the users and, wherever possible, to keep experimental conditions constant. "Repeated measures" designs should generally be employed, because of the small number of users available. Obviously, these practical difficulties give rise to missing data problems, resulting from incomplete conditions, cause the loss of levels and factors from designs, and make the systematic varying of conditions in pilot studies difficult.

In addition, the increased range and skewed variability resulting from the range of motor deficits, leads to increased noise and violation of the assumptions of many statistical tests, such as normality of distribution or homogeneity of variance. Where statistical tests are possible without violation of standard assumptions, they should be carried out. However, even if the power of these experiments was unknown because of the reasons outlined and the small sample size, the effect sizes may still be large because of the sometimes radically different behaviors that are associated with different functional impairments. For this reason, some statistical results that are significant do not appear so, but should be analyzed in terms of statistical power ($1 - \beta$: the probability of rejecting a false null hypothesis—see Cohen, 1988), and estimates of effect size given (Chin, 2000).

If all else fails, the experimental evidence should be presented as primarily qualitative, allowing the experiments to be used as pilots for retrospectively estimating effect sizes and power, enabling a continued effort to increase the effectiveness of the experimental series. At the very least, the researcher can use the argument that "*this* user had *this* (very real) difficulty." While the immediate response to an assertion like this is that it contravenes most accepted scientific principles, when dealing with real people in the absence of large sample groups, this very pragmatic approach may be persuasive to the commissioner of the research. This is especially true if the commissioners of the design subscribe to the "top-down" design principle whereby if the design can be used by the least able users, then the design should also benefit more able users (Keates and Clarkson, 2003).

Defining "Acceptable" Results

An "acceptable" level of accessibility can be defined in many ways. Perhaps the most persuasive is to interpret it in terms of overall productivity of the user. Productivity is, in turn, often expressed in terms of minimizing the time to complete a task, often combined with keeping errors and unnecessary (i.e., repetitive or redundant) actions to a minimum.

It is often argued that for a product to be "acceptably accessible" it should represent no differentiation between users solely because of their functional capabilities and, furthermore, that the overall productivity of each user should be identical. In other words, a user with a disability should be able to complete a task with as few errors, and in the same amount of time, as someone who is nominally able-bodied. After all, it is no good claiming that a product is accessible based principally on whether someone can complete a task. For example, a task that takes an able-bodied only five minutes to complete using the product, but takes someone who is blind five hours to complete cannot realistically be considered acceptably accessible.

Thus, a simplification of this argument is that all users should be able to complete a specified task using the product in the same amount of time. Any product that fails to meet this target is deficient in some way. However, this argument is an oversimplification of the issue.

For example, while the goal of equal time to complete a task may be the "ideal," it is likely to be of little use when considering whether the level of accessibility provided by a particular product is acceptable. The reason for this is that it is unlikely that the same time to complete a task is achievable for people with disabilities as for nominally able-bodied users, unless the task is a very simple (i.e., straightforward) one. Someone with a visual impairment is unlikely to be able to perform a visual task rendered aurally in the same time as a sighted person could accomplish the task. Similar levels of task completion times may be possible for a highly sequential task (e.g., listening and answering a linear series of questions), but not for more randomly organized tasks (such as seek-and-find).

Studies comparing the time taken for users with motor impairments to complete a task compared with the time for "able-bodied" users have shown very few instances where the users with motor impairments matched the performance of their able-bodied counterparts (e.g., Keates et al., 2002). In those instances where similar performance is observed, they were usually when the performance of the able-bodied users was being compromised in some way. Some of those tasks were simple comparisons of unmodified systems and so such differences would be expected. However, many of the studies were based on tests involving systems that had been modified to compensate for the impairments and so should, in theory, have minimized or mitigated the effects of the particular impairments being investigated.

Additionally, it has to be remembered that any useful compensation that can be offered to users with motor impairments usually also benefits the able-bodied users. Thus, while it may be possible for a particular form of assistance to improve the overall time to complete a task (or whatever metric is being used to measure "productivity") for a user with a motor impairment, it is likely that the time for an "able-bodied" user would also improve when using the same assistance. In other words, the goalposts are prone to moving.

An example of how different users behave fundamentally differently can be found in a simple set of experiments based around a basic cognitive model, the Model Human Processor (MHP—Card, Moran, and Newell, 1983). The Model Human Processor states that the time to complete a task is described by Equation 10–1.

CHAPTER TEN

$$\text{Time to complete} = [x \times T_{(p)}] + [y \times T_{(c)}] + [z \times T_{(m)}] \qquad (10\text{--}1)$$

where:

- $T_{(p)}$ is the time for one perceptual cycle, i.e., the time required for the user's senses to recognize that they have sensed something.
- $T_{(c)}$ is the time to complete one cognitive cycle, i.e., the very simplest decision-making process in the brain. This can be thought of as being effectively akin to a "yes/no" question, such as "did I see something?" which would take one cognitive cycle to answer. A more complicated question would take several cognitive steps because it needs to be broken down into a series of either "yes/no" questions or other simple classification steps.
- $T_{(m)}$ is the time to produce a single, simple motor movement, e.g., a downward stab of a finger onto a key. Releasing the key would require a second motor movement.
- Finally, x, y and z are all integer coefficients (1, 2, 3, etc.) that represent how many of each component cycle are required.

Card, Moran, and Newell (1983) describe a number of basic tests that can be performed to calibrate this model. Although the tests they describe are paper-based, they are straightforward to implement within a computer program. Such a program can be used to screen potential user trial participants as the results it generates provide a useful indication of the level of severity of motor impairment of a particular user. For an able-bodied person, the typical observed times of each cycle from such a program are as follows (Keates et al., 2002):

$$T_{(p)} = 80 \text{ ms}$$
$$T_{(c)} = 90 \text{ ms}$$
$$T_{(m)} = 70 \text{ ms}$$

For users with moderate to severe motor impairment, the observed $T_{(p)}$ is typically 100 ms—close to the 80 ms seen for the able-bodied users, but still 25 percent slower. The observed $T_{(c)}$ for the same users is usually 110 ms—compared with 90 ms for the able-bodied users. Note that even though the observed times are somewhat slower for the users with motor impairments, they are still close enough that the differences are not particularly important when considering the interaction as a whole.

However, the results for $T_{(m)}$ is quite complicated and provide an informative insight into how functional impairments can fundamentally change a person's behavior. The observed results for users with motor impairments typically fall into one of 3 bands: 100–110 ms, 200–210 ms, or 300–310 ms—all of which are significantly slower than the 70 ms observed for the able-bodied users.

The reason for the different bands is suggested by the 100 ms time differences between them. That difference is the same magnitude as either additional $T_{(c)}$ cycles or $T_{(m)}$ steps. If it is additional $T_{(c)}$ cycles that are present, then that implies that more cognitive effort is required of the user to perform each "simple" motor function than for able-bodied users. The likely cause of the $T_{(c)}$ cycles is the extra cognitive effort required to produce carefully controlled movements, for example, by having to try to suppress spasms or tremor. This extra cognitive effort has a knock-on implication that so-called "automatic" responses are not achievable for those particular users. Alternatively, if additional $T_{(m)}$ steps are present then that implies any supposedly "simple" action, such as pressing down on a key, is not performed as a single movement, but actually several (usually) smaller ones.

Irrespective of the cause of the different bandings, the net effect is that some users simply cannot perform basic physical actions as quickly as other users can. The data presented here has been derived from users with motor impairments. Similar results are also seen in much of the literature on ageing research, showing that, for example, increasing age is strongly correlated with slower response times on simple reaction tasks. However, in the case of ageing in particular, older adults are often able to compensate to some degree for their increasingly slower response times through their increased experience and knowledge, acquired over the years. A direct consequence of this user variability in the performance of "simple" tasks is that trying to design a product to meet the requirement that all users be able to complete a task in the same time, whether nominally able-bodied, older or functionally impaired, would be almost impossible for many products.

Thus the most pragmatic definition of an acceptable level of accessibility for a particular product is one that allows all user groups to perform a series of tasks within an acceptable level of tolerance on an agreed range of metrics. The metrics will most likely be focused around productivity measures, such as time to complete a task, throughput and error rates. It is also likely that there will be no single, universal tolerance that can be implemented across the entire range of a company's products. Instead, the best that can probably be considered is a framework for defining what levels of tolerance are acceptable for the particular product being evaluated.

Such a framework would most likely start off by performing a step-by-step task analysis, breaking down the interaction into its component steps. This should be done for each of the user groups being considered—including the role of the particular assistive technologies that are required to enable access. For example, if someone using a screen-reader needs to perform 50 actions to complete a task, whereas everyone else need only do five actions, then clearly there is a problem with the model of interaction for the use of a screen-reader. Under such circumstances, the onus should be placed squarely on the shoulders of the product designers to explain why their interaction model for users who are blind appears to be so poor.

The framework should also contain recognition of the level of severity of the person's impairment. Someone with a minor tremor will most likely be able to complete a task faster than someone with a frequent major spasm. No assistive technology exists at the moment that could level that particular playing field. Equally, the framework would also need to look at error rates. Errors are a major source of delay and increased time on task for all user groups, but especially older adults and those with disabilities. If a person's rate of interaction with a product is already slowed by the presence of functional impairments, then it is imperative that errors be kept to an absolute minimum. Continual errors affect the user's perception of a product and

TABLE 10–1
A Possible Framework for Deciding Whether a Product Is "Acceptably Accessible"

Stage	Action
1	▪ Identify each target user group (or persona) to be considered
2	▪ Identify component steps in the interaction for each target user group and for the base-line, datum user group (most likely younger, able-bodied users)
3	▪ Compare number of steps for each group
DECISION GATEWAY—1	Are there the same or very similar numbers of steps for each user group? ▪ *If not—significant differences have to be justified to senior management or remedied.*
4	▪ Perform user studies with the baseline user group—calculate times to complete tasks and sub-tasks and error rates
5	▪ Perform user studies with each target user group—calculate times to complete tasks and sub-tasks and error rates
DECISION GATEWAY—2	Could all of the target users complete the tasks? ▪ *If not—the causes of difficulty need to be removed or redesigned and remedied.*
6	▪ Compare the error rates for each group
DECISION GATEWAY—3	Are the error rates the same or very similar between the target user groups and the baseline group? ▪ *If not—significant differences have to be justified or remedied. (Note: this should be extremely hard to justify)*
7	▪ Compare the times to complete tasks and sub-tasks modified by the following factors: a. the number of component steps (if it has been agreed that notable differences are OK in this case) b. the proportion of component steps affected by each group's particular disability (sight, hearing, etc.) c. the relative "importance" of the step (high, medium, low) d. the level of severity of the disability (high, medium, low) [Note—this is likely to be an inverse factor] e. any additional latencies (whether innate in the user, or arising from the AT used, e.g., the time taken for a screen-reader to read out the text on-screen)
DECISION GATEWAY—4	Are the modified times to complete tasks and sub-tasks the same or very similar? ▪ *If not—significant differences have to be justified or remedied. (Note: this should also be extremely hard to justify after making the modifications to the times listed in stage 7 above)*

Involving Users in the Design Process **127**

thus the willingness to keep using the product diminishes. In addition, the time to correct any errors is increased for these users, and so they are, in effect, paying a double penalty for errors. Therefore, in summary, a framework for deciding whether a product is acceptably accessible could look something like Table 10–1.

An example of modifications to times to complete tasks would be an estimation of all the different latencies that are present in the system. For instance, if there is a 5-second delay in generating the speech output for a screen reader and there are 10 such outputs, then a total latency of 50 seconds should be factored in to the tolerance calculation. If a sighted person can read at (say) 150 words per minute, but a blind person can only listen at (say) 100 words per minute, then a tolerance of 3/2 should be factored in. This framework recognizes that directly comparable performance is unlikely because of a whole range of factors, but tries to identify what should be achievable in a more constrained fashion than simply "target users can or cannot complete the tasks."

SUMMARY

This chapter has discussed many of the issues to be considered when involving users in research and design activities. While many of the widely accepted methods of performing user studies often require large homogeneous sample sets, this is not always possible when considering design for universal access. In real, practical circumstances where there is limited time, money, and user availability, researchers and designers may be required to tailor their data collection methods to meet those constraints. It is particularly important, therefore, that any user involvement is designed carefully, to ensure that the maximum amount of useful information is obtained and then packaged in the most effective manner. Researchers, who typically obtain this data, and designers, who typically use it, need to work together to ensure that these goals are achieved.

KEY POINTS

- Involving users in the design process is a requirement of user-centered design.
- Users should be selected accordingly to carefully considered criteria.
- All user research methods adopted should be adapted to meet the particular circumstances of designing for accessibility and not applied without suitable forethought and planning.

11

Conducting Sessions
With Users

The previous chapter discussed the issues underlying the involvement of users in the design process. In this chapter, we will examine the practical implications of conducting user trials with users with a range of capabilities and impairments. For our purposes here, we will assume that the company, research or design team wishing to perform the user trial sessions already has a framework in place for finding and recruiting users. When conducting a user trial session, the main stages to consider are as follows:

- pre-session preparation;
- selecting and preparing the trial location;
- conducting the sessions; and,
- post-session follow-up.

We will examine each of these stages in detail.

PRE-SESSION PREPARATION

Whenever considering conducting user trial sessions, it is imperative that the trials undergo a comprehensive planning stage in which as many possible eventualities are identified and solutions put in place. The time spent with users is a very valuable resource and it should not be squandered by having to waste time attending to inconsequential details that should have been taken care of beforehand.

Planning the Tasks to Be Done

The central part of planning for any user session has to be deciding what tasks are to be performed during the session (Nielsen, 1993). Typically, for a usability study, the tasks would be focused on ensuring that the overall functionality of the product was straightforward for the user to use. Thus, tasks would be based around mimicking what end-users would be expected to do with the product. In the case of designing for accessibility, the goals of the traditional usability studies still need to be met and so it may nevertheless be necessary to conduct "usability" sessions as well (if they cannot be incorporated into the "accessibility" sessions). Designing a product that is not usable is simply not an option.

However, in addition to those goals, specific tasks focused on the accessibility of the product may also need to be included. For example, if assessing a new peanut butter jar, a usability task may be "Make a peanut butter sandwich," whereas an accessibility task may be "Read the ingredients label aloud," a task that few usability studies would be likely to address. Note that if the users can read the entire label, but cannot make a peanut butter sandwich (implying, for example, an inability to identify the jar as containing peanut butter or being able to open it), then the product is still useless as far as the user is concerned. This just reinforces the fact that the precise distinction between usability and accessibility is largely a semantic one, albeit also an important one.

Getting Ethical Approval

Whenever conducting experimental sessions of any type that involve members of the general public, it is essential that some kind of ethical review procedure is followed. For many academic researchers, their university should have an ethical review board that ought to be consulted. In most U.S. universities, the Institutional Review Board (IRB) is the body that would typically oversee this process. Companies often do not have such bodies overseeing their work, and so should consider establishing an internal body that serves a similar purpose. Particular issues that need to be addressed include whether the experimental protocol:

- makes it clear that the participants can pause the session at any point;
- makes it clear that the participants are free to withdraw from the session at any point;
- makes it clear that the participants can withdraw their data from the post-session analysis at any point;
- outlines the data to be collected, how it will be handled, and who will be able to access it;
- outlines the process by which the data will be fully anonymized;
- describes the risks to the participants in clear and unequivocal language;
- provides examples of consent forms that participants will be expected to sign; and,
- ensures that at no point are the participants exposed to potential risks of "harm" (not just physical, but also emotional).

This list is not complete, but gives a flavor of the issues that need to be considered. Many IRB forms are multiple pages in length and have increased in volume significantly in recent years. Also note that getting ethical approval can be a time-consuming business, so this process should be started as far in advance of the planned user sessions as possible.

It must be noted that ethical approval can be particularly difficult to obtain when considering users with cognitive or learning difficulties. Review boards may take the stance that such users may not necessarily be able to understand what is involved in the trials and therefore whatever they are being asked to agree to. Provision will most likely have to be made for getting informed consent from a person's parents, legal guardian, attorney, or other designee. A number of universities and medical schools have had success in getting their IRBs to agree to studies held with such participants. It may be helpful to cite such approvals in applications for ethical approval and offer to follow any provisions required under the terms of those approvals.

Identifying the Needs of the Participants

Once the users have been identified (using the techniques described in the previous chapter), it is worth spending time becoming familiar with who they are, especially their particular needs (Gheerawo and Lebbon, 2002). A list should be drawn up of each of the users and their particular requirements, especially in terms of assistive technology. If a person who is blind, for instance, is participating in a trial session focusing on a particular computer program, then it would be helpful to know if that person used a specific screen-reader package. Similarly, if the participant is deaf and requires a sign-language interpreter to be present, space needs to be provided for an extra person. The principal areas to focus on are:

- **Navigation:** How is the person going to get to the trial location and navigate around the trial venue?
- **Communication:** What mode of communication should be used (written, spoken, Braille, sign language, etc.)?
- **Assistive Technologies:** Will the participant need access to specific assistive technologies during the user trial sessions and are those assistive technologies available? If they are not available, can the participant bring his or her own?

Conducting Pilot Studies

Pilot studies are an invaluable option for debugging user study protocols and no accessibility user trials should be attempted until a pilot session has been held. Even experienced accessibility designers and practitioners will find something that they have overlooked.

A pilot study does not necessarily need to involve real users, but ideally they should. It does, however, need to involve someone walking through the complete session as a user to make sure that all of the technology is working and that all of the wrinkles in the protocol (e.g., repeated questions, questions that do not make sense, impossible tasks) are ironed out. Stumbling around trying to figure out what to do next in front of real participants is neither dignified nor professional. Remember, practice makes perfect.

In particular, when considering design for accessibility, it is important to ensure that the product being tested is compatible with whatever assistive technology is expected to be used by the participants. If the product is not compatible, then the sessions will be almost entirely unproductive. Note that trying to force or coerce someone into attempting a task for which they would normally expect to use a particular assistive technology while denying them access to that technology (as would be the case if it was incompatible) would almost certainly be considered unethical behavior.

Preparing and Distributing Pre-Session Documentation

It is important to send as much documentation as possible to the participants ahead of the sessions. In this way, users who need extra time to read through and understand the documentation will not feel pressured to do so while being watched (as they may well view it) by the session organizers. The documentation that should be sent out includes copies of:

- all consent forms that the participants will have to sign;
- details of any reimbursement offered;
- details of the time and location of the trials;
- directions to the location where the trials are to be held; and,
- the tasks to be performed and details of how the session will be structured and run (the experimental protocol).

Note that the format and content of the documentation sent out will need to be tailored to meet the needs of each individual participant. For example, participants who are blind will need any documentation to be provided either in Braille or in an audio format (e.g., on tape), based on their particular needs. Participants with low vision will need versions with larger font types. It is advisable to ask such users what size font they can read before printing out the documents.

Users with motor impairments may possibly have difficulty handling paper documents. In this and in all of the cases above, making electronic versions of the information available, such as through a Web site, may also be helpful (as long as the web pages are themselves accessible).

Participants who are hard of hearing should be able to manage with standard text documents, but those who are deaf may have difficulty because most forms of sign language have a different grammatical structure to typical spoken or written languages. In this case, the best solution is probably to use very simple, clear language. Similarly, for users with cognitive or learning difficulties, keeping the language clear and simple is also the best option. For specific learning difficulties, such as dyslexia, providing audio versions of the documentation may also help.

Finally, the background and level of knowledge of each user should be considered. Describing something to a person in their teens or someone in their eighties may require a different set of metaphors, analogies, and jargon. Take the word "wire-less," for instance. To people over a certain age this word is probably synonymous with a radio, whereas for many others it simply means something that is portable. Thus, a "wireless network" could mean radically different things to different participants. Consequently, it is best to play it safe and either avoid all use of jargon or else include explanations of all terms that the participants may not understand.

The golden rule is: do not make assumptions about the needs of the participants. If in doubt about what format they need the documentation in, ask them what they would prefer.

SELECTING AND PREPARING THE TRIAL LOCATION

The first choice faced when considering where to hold the user trial sessions is whether they should be held in a facility specially designed for such sessions, such as a usability laboratory, or else at another location. Other candidate locations would most likely include ones that are more familiar to the participants, for example a day-care center. Participants' homes should only be used as a last resort because of potential liability issues, such as what happens if a personal possession gets broken during the course of the session.

Usability Laboratory

Holding trials in a usability laboratory has many advantages. For a start, if all of the user sessions are held in the same laboratory, this means that the team conducting the sessions can ensure that the conditions in which the trials are held are the same for each participant. For example, it is extremely unlikely that any interruptions will happen, such as when someone walks in by mistake, whereas that kind of thing does happen in other locations. This continuity is very helpful in that it allows more direct comparisons between different users, as fewer environmental variables have to be taken into consideration. Anything that reduces variability between user sessions, other than the variability that arises from the users themselves, is highly desirable.

Usability laboratories are often designed either with all the necessary monitoring equipment (video cameras, tape recorders, etc.) already built in, or at least space for the equipment to be installed. Monitoring equipment is important because it allows the designers and human factors experts to analyze the sessions in detail, once they have been completed. In addition, when using a usability laboratory, the team conducting the sessions should have plenty of time to find the best possible arrangement for all the equipment, both for monitoring the sessions and also for the product being tested, plus any additional assistive technology as required.

There are disadvantages to using a usability laboratory, though. It will almost certainly be an unfamiliar place for the participants and a potentially disorienting one. As such, the results from the user sessions may not reflect accurately the real-world use of the product being investigated, as the participants have to adjust to being in the laboratory while also attempting to perform the tasks. Having said that, virtually all usability trials are artificial in some way (e.g., the users are often told the tasks that they need to accomplish rather than deciding them for themselves as would be the case in the real world), so this is not necessarily a major problem.

Other problems that may arise include the issue of how the participants have to get to the usability laboratory. If the users are not particularly mobile, this may present a significant logistical challenge. It may be easier to take the assessment team to the users, rather than the other way round. If some of the users do try to make it to the trial-site, then they need to be provided with directions for getting there in a format that is accessible for them (see the earlier discussion on preparing and distributing pre-session documentation).

The navigation challenges do not stop once they reach the trial-site. Imagine, for instance, being blind and being dropped off outside a large building, having been told that somewhere in that building is the laboratory that you are searching for. Clearly, thought needs to be given to how participants navigate within the trial-site to reach the usability laboratory. One option is to ensure that someone is waiting for the users and escorts them to and from the laboratory. Another option is to ensure that the laboratory is well signposted throughout the entire building (including Braille labeling). Note that some participants may deviate from the preferred route by accident. Consequently, if the signposts are only on the preferred route, then those participants could end up well and truly lost. Clearly, having someone escort the participants is the better option, but only if someone is available to fulfill that role. Most of these considerations are not new for building managers. They will most likely have already been considered when making sure that a building or facility is fully complaint with, for example, the Americans with Disabilities Act (ADA, 1990) and appropriate buildings accessibility standards, such as BS 8300:2001 (BSI, 2001).

Another challenge when conducting sessions in a usability laboratory is that if the participants are used to a specially adapted environment when using the type of product that is being tested, then that environment needs to be replicated in the laboratory. This can be a time-consuming business and may not match the participant's own setup in every detail. In addition, if the participant is used to using a particular set of assistive technologies and those are not available within the laboratory, then the participant either needs to bring in those assistive technologies, or else try to manage with substitute equipment. Such difficulties arise when the participants are used to using custom-built devices designed or tailored specifically for their needs. In summary, Table 11–1 shows some of the particular challenges faced by participants.

Note that all of the users in Table 11–1 may be used to using specific assistive technologies. Table 11–1 is not a complete list of all of the needs of individual participants. Again, if in doubt about what assistance a participant may need—ask them.

Other Locations

User trial sessions do not always have to be held in a usability laboratory. It is possible that they could be held in the environment where the product being investigated

TABLE 11-1
Examples of the Challenges and Needs of Accessibility User Trial Participants

Participants Who Are:	Challenges and Needs:
Blind	■ Navigating to and around the trial-site (e.g., to the rest-rooms) ■ Getting used to the layout of the laboratory ■ Provision for a seeing eye dog to accompany the user (e.g., space for the dog and access to drinking water, etc.)
Low vision	■ Large, clear signposting
Deaf	■ Provision for a sign-language interpreter
Motor impaired	■ Access to the building (e.g., wheelchair access, not too many steps, etc.)

is to be used (in an office or factory, for example) or else somewhere that is more convenient for the participants. For example, if several participants have been recruited for a local day-care center, then it may make sense to move the user trials to there. For our purposes here, we will assume that a day-care center is the alternative location being considered for the trials.

There are several advantages to holding the trials in such a venue. For a start, all of the participants from the center should know how to get there and have their transportation arrangements already in place. They should also know their way around the building and so the need for signposting is removed, or at least reduced. Additionally, the individual members of staff of such venues are often very helpful and usually willing to find and coordinate the users, making sure that they all turn up to the correct room on time.

The environment in a day-care center is much more likely to be relaxed and more "natural" than in a usability laboratory. As such, more of the true character, feelings, and thoughts of the participants may show through in the trial sessions and this, in turn, means that the data obtained should be more "realistic." It is also likely that any assistive technologies required by the users will already be present in the center. For instance, many people who attend day-care centers often take advantage of computer-training courses to use the computers there. As such, the centers frequently provide the assistive technologies required to enable them to access the computers, or else the users bring in their own devices. Consequently, if the product being investigated is computer-based, then it is likely that the assistive technologies required by the participants will be readily available.

The final advantage of holding the user trial sessions in somewhere like a day-care center is that other potential participants may become interested in what is going on. They might then volunteer to take part in the study or else in future ones. Having keen volunteers willing to take part in such trials is always a good thing.

Needless to say, there are also drawbacks to hosting user trial sessions in a day-care center. The main drawback is the lack of control over the environment faced by the team organizing and conducting the sessions. For example, the team will most likely have to settle for whatever room is available in the center on the day of the trials and the room may be either too small or too big. Additionally, the room may not be soundproofed particularly well meaning that other activities in the center may impinge on the trials. Equally, any noise from the trials may disturb the other activities and it is important to keep all disruption of the day center's normal business to a minimum if the team is to be invited back. It is important to be sensitive to the needs of the day center, especially after all the time and effort undoubtedly put in to find users and set up an agreement with the center. Being kicked out for being too disruptive would not be a good thing.

Another drawback is the potential for interruptions and other distractions. Staff at the care center may reserve the right to enter the sessions at any point to ensure that the participants are not becoming unduly stressed. Equally, sometimes other visitors may walk in simply by accident. The team conducting the trial session should be prepared for such interruptions. There are also logistical challenges to holding sessions off-site for the team. All of the equipment required for the trials needs to be packed up, transported to the center, unpacked and set-up, then re-packed and transported back to the company's site. Sometimes the set-up of the room will not allow for the optimal placement of monitoring equipment and so compromises may have to be reached. Another important consideration is that the room may not be secure, in which case equipment should not be left unattended, especially over lunchtime. Many of these logistical challenges can be overcome though, with a bit of forethought. Indeed, when considering both types of possible venue, i.e., usability laboratory compared with other locations, none of the drawbacks are particularly difficult to overcome or make allowances for. What they do require is some planning ahead and thinking around possible solutions.

Remote Locations

There is a third option for the location of the user trials, namely remote locations. The assumption for the "other locations" described above was that they were within a practical distance of the team conducting the sessions. However, not all potential participants are local to the team and it may occasionally be necessary to perform the trials remotely. Needless to say, remote user trials are the least preferred option as far as the team is concerned and also, arguably for the participants. Their principal advantage is that they offer an opportunity to include participants who might not otherwise be able to take part in the trials and allow the team to gather data from them.

Of course, the main drawback is that the channels of communication with the participant and observation of the trial session are limited. The most common method of conducting such trials is over the telephone, but this obviously causes difficulty when working with participants who are deaf, although there are technological solutions to this such as the use of text telephony (TTY) machines or even instant messaging systems. Perhaps a less obvious difficulty arises when trying to explain concepts that are new to the participant or that the participant is having particular difficulty understanding. When trying to explain something face-to-face, the person doing the explaining can employ a range of methods of communication, such as gestures or sketches, but these are often not possible when communicating remotely with someone.

Perhaps the product family that is most suited to remote trials is computer software. It is possible to write logging software that records the entire user interaction with the software as well as the specifications of the computer on which the software is being run for later analysis. The data that is logged can be transmitted via a broadband connection in real-time or else stored in a file that can be e-mailed to the team later. Even so, there are still significant logistical problems to be overcome. The software needs to be sent to the participants, either by e-mail or else on a CD or DVD through the mail. The participants then need to know how to install and run it. If the data-logging file needs to be e-mailed back to the team, then the participants need to be able to find the files and know how to attach them to an e-mail. These may sound like straightforward tasks for most people, but if designing software for people who are having difficulty accessing existing programs, then it is likely that at least some participants will have difficulty accomplishing these tasks. Additionally, such users are also more likely not to have broadband connections and, consequently, are more likely to need to have to do these more complicated tasks.

CONDUCTING THE SESSIONS

A typical session would consist of four main components.

1. pre-trial briefing;
2. use of the product/performance of the tasks;
3. interviews; and, finally,
4. post-session debrief.

As with all usability testing, the participating users need to be treated with respect and courtesy at all times. This includes being aware of the participants' willingness, or otherwise, to discuss their particular impairments or difficulties. Some participants will be open and frank when discussing themselves, others will be more reticent. If a participant does not appear to be willing to talk about such topics, then do not attempt to coerce them into doing so. This would constitute unethical behavior.

Pre-trial Briefing

During the pre-trial briefing the participant is usually asked to review any documentation sent out prior to the session, including consent forms and descriptions of what the session will entail. Needless to say, the issues of format and content of the documentation sent out prior to the sessions also apply here. Additionally, if the participant is either deaf or hard of hearing then the person conducting the briefing has to be prepared to conduct the briefing either through written notes, or else through a sign-language interpreter. It is also essential to remind the participants that they are free to retire from the study or withdraw any data collected about them, at any point, and that this will not affect any reimbursement that they may have been offered for their participation.

The person conducting the session should talk through and explain each piece of documentation to ensure that the participant has understood fully what he or she is being asked to do. The participant should then be offered an opportunity to ask whatever questions the participant needs to clarify anything he or she may have been unclear about. The final stage of the pre-trial briefing involves asking the user to sign the consent forms to begin the session.

Performance of the Tasks

When dealing with users with more severe impairments, it is especially important to be sensitive to their physical needs. For example, such users will often tire more easily than the person conducting the session may normally expect. The comfort and well-being of the participants must take priority at all times. This takes precedence over the defined experimental protocol. Participants should not be coerced into attempting tasks that may be simply too much for them, either physically or emotionally.

Thus, if it becomes clear that a particular task is proving too difficult or frustrating for a participant, then another means of completing the task should be offered (e.g., using the keyboard instead of a mouse for a computer-based task) or else that bit of the task abandoned. Similarly, if a participant is becoming noticeably tired, or reluctant to continue, then they should be reminded that they do have the option of taking a few minutes break, skipping the more demanding tasks or even retiring from the session at that point. It is also important to keep reassuring the participants. Older participants, in particular, seem to worry that somehow they will show their ignorance of something, especially something they regard as "high-tech." They are also often very nervous about breaking the product, again especially if it is high-tech.[1] Participants who are worried about such matters are clearly not going to behave as they might in more relaxed surroundings.

Likewise, the number of observers should be kept to an absolute minimum. It is quite common for several members of a design team to feel that they should be present at the trials, but having too many people in the room will unsettle the participants, especially those who are already nervous. The ideal arrangement is to have one person present to conduct the trial, guide each participant and so on, while a second person attends to the monitoring equipment. If more people feel that they absolutely have to watch the session firsthand, then a remote video feed to another room (well out of earshot of the participant) should be set-up and they should watch from there.

A common technique adopted during user trials is the "think aloud" approach. This technique requires the participants effectively providing a narration to what

---

1. I have often witnessed comments such as "You will promise me that this is not going to go up in a puff of smoke?" or "You won't blame me if I break it, will you?"

they are thinking and doing. It can provide very useful insights and data for later post-session analysis. However, think aloud is not suitable or appropriate for all users. Clearly users who have difficulty speaking, for whatever reason, will not be able to perform a think aloud task. Additionally, users with more severe impairments may find the demands of thinking aloud (and it *is* demanding—just imagine trying, for instance, to talk through every step and decision when making toast) combined with the demands of performing the tasks simply too much. In such circumstances, interviewing the participants after they have attempted or completed each task is a better option.

One of the other major issues to consider when examining how the participants perform the tasks is the presence of coping strategies (DTI, 2000). Many people with functional impairments find strategies for compensating for their impairments— sliding heavy objects that were designed to be lifted, using two hands instead of one, making customized alterations to products to make them easier to use. Identifying coping strategies can be difficult for someone who is not familiar with the nature of functional impairments. Users will often perform the coping strategy as if it was second nature to do so (through practice) or, alternatively, may actively disguise any such strategies to avoid drawing attention to any functional restrictions that they may have.

Even when coping strategies have been identified, finding the cause of them is not always straightforward. For example, performing a one-handed operation with two hands may be a coping strategy for maneuvering an object that is too heavy, but it is also a strategy for increasing accuracy. Note that if coping strategies are observed, they are not a failure of the study. While they may make post-session analysis of the data more complicated for the design team, their presence in the trial provides a useful insight into how real users may try to use the product being investigated. They also offer a useful alternative view from that of the design team of how the same task could be accomplished.

Interviews and Final Debrief

Interviews differ from the final debrief in that they are interspersed throughout the tasks, whereas the debriefing occurs only at the end. As such, interviews not only offer an opportunity for the team to gather important information from the participants about their views on how the trial is progressing, such as any difficulties completing the tasks, but also an opportunity for a rest. Interviews allow the participants time to recuperate between tasks, while still providing useful information to the team, rather than, say, just sitting around for a few minutes.

For the final debrief, it is important to remind the participants that they are free to withdraw their data at any time. If any further data is expected to be needed after the session then the debrief is the best time to ask for permission to contact the participant afterwards. As with the opening pre-trial briefing, the person conducting the interviews and the final debrief needs to be prepared to conduct the interviews through pen and paper or through a sign-language interpreter if the participant is deaf or hard of hearing.

POST-SESSION FOLLOW-UP

Whenever conducting a user study, it always pays to be polite, especially if it is likely that the participants may be asked to take part in future studies. This is particularly likely in the case of designing for accessibility because:

- Suitable users are that much more difficult to find than for usability studies; and,
- Companies tend to design families of products of a similar nature and so participants who are suitable for inclusion for one product are quite likely to be suitable for other members of the product families.

Consequently, it is worth sending a thank you note to the participants (in an appropriate format—text, audio, Braille, etc.) for their time and effort. After a couple of weeks have passed, it is also often worth asking users to think again about their experiences when using the product and completing the tasks. During the trials, their attention may have been diverted by all of the distractions, the video cameras, all the people watching them and so on.

SUMMARY

Conducting user trials is a cornerstone of designing for accessibility (Keates and Clarkson, 2003). It is very difficult, arguably impossible, even for experts to design a genuinely accessible product without some type of user trial. User trials are, though, comparatively expensive and time-consuming. Therefore, it is necessary to design the trials to get the maximum useful information from the participants in the minimum amount of time. The structure of the trials needs to be a combination of robustness and flexibility—robust enough to accommodate as much variability in user capabilities as possible, flexible enough to allow adaptation of the tasks when the user capabilities simply cannot be accommodated in the basic design.

Do not be disheartened if your first attempts at an accessibility trial do not go according to plan. It takes practice to get used to designing such trials. This reinforces the notion of how important it is to perform pilot studies, which can be used to identify the most serious failings and offer an opportunity to amend those features of the trial protocol.

The next chapter is a case study of an accessibility study into the design of digital terrestrial television and illustrates many of the issues raised in this chapter and in Chapter 10.

KEY POINTS

- Do not make assumptions about the needs of the participants. If in doubt—ask.
- Make sure that the most relevant users are selected to participate in the trials.
- Plan ahead and be prepared for unexpected results.

12

Case Study—Investigating the Accessibility of Digital Television for Older Adults

Throughout this book, we have examined the importance of putting users at the center of the design process. Equally important is the issue of identifying the correct users with whom to work. In Chapters 10 and 11 we looked at the specific detailed issues of working with users and the particular issues involved when working with users who would not typically be included in a user study for a "mainstream" product, but who do need to be involved when designing for accessibility.

The case study presented in this, the final chapter in this book, illustrates many of the principles discussed being applied in practice. It looks at the design of a new digital television service being offered in the UK and whether the design of that service was accessible for significant sections of the viewing audience, especially older adults. The user studies performed to assess the new technology are described along with the somewhat startling results obtained, and examples of how the results were packaged for easier consumption by designers.

This study was conducted in the summer of 2003. The net effect of this study was that many of the aspects of the service that were highlighted as not offering acceptable levels of accessibility and presenting unnecessary levels of difficulty have since been comprehensively re-designed by both the manufacturers of the set-top boxes and the broadcasters. As a direct result of a fairly small-scale and comparatively cheap set of user trials, notable improvements in the accessibility of a nationwide service have been implemented. This study shows that a well-designed set of trials can be highly influential and of significant benefit to many people.

INTRODUCTION TO THIS USER STUDY

The aim of this study was to investigate the accessibility of digital television (DTV) technology available in the UK, focusing on the current generation of set-top boxes

(STBs) which provide "free-to-view" services. It was commissioned by the UK Department of Trade and Industry (DTI) and its objective was to identify specific causes of concern with regard to user interaction with DTV that might lead to exclusion, i.e., situations where users may be unable to use the new technology. Of particular interest was the accessibility of the technology for older adults, not only in terms of physical access, but also in terms of how well they were able to understand and predict the system's behavior. There was a particular focus on older adults as they are the largest user group that may be expected to experience difficulties with the transition from analogue to digital television.

The UK Television Market

In the UK, the majority of domestic households receive television through terrestrial broadcast station signals arriving via an aerial, often mounted on the roof of the house. A smaller proportion of households receive televisions via satellite broadcast or through cable networks.

Starting in 2008 and scheduled to be completed by 2012, the UK Government aims to switch over completely from analogue broadcasting to digital only services. Following this switchover, analogue television services will no longer be available within the UK. Thus the Government has realized that it is essential that the supporting digital technologies must be accessible for all. After all, no government wants to be accused of denying access to a service as ubiquitous as television. In the meantime, broadcasters are transmitting both types of signal at additional expense as both types of network need to be maintained. The broadcasters are understandably keen to perform the switchover as soon as possible and, by extension, so is the UK Government, because the largest broadcaster (the BBC) is funded exclusively through license fees.

The Differences Between Digital and Analogue Television

The majority of televisions in the UK are fundamentally analogue in nature. To display a digital signal (consisting of a series of discrete 0s and 1s) on an analogue television, the signal needs to be converted to analogue (i.e., a continuous waveform). Some televisions had the electronic circuitry to do this conversion internally and these are referred to as integrated digital televisions (iDTVs). However, for all televisions without this circuitry, an external box is needed to perform the conversion. These are usually called set-top boxes (STBs) and are expected to be the most common method of viewing digital television in the UK in the short- to medium-term, because they are much cheaper to buy than integrated digital televisions. In the longer term, iDTVs are expected to dominate through the natural product replacement cycle as older analogue-only televisions wear out and owners decide to replace them with iDTVs.

Consequently, viewers of digital television will usually have two separate components with which to interact: the television itself and the STB, each of which comes with its own remote control. This represents an immediate increase in the level of complexity of the interaction for the users. While in theory it should be possible to control both components through the STB remote control, in practice it is often the case that some functions, such as the volume control, do not always work. Thus it is often necessary to use both sets of controls.

Finally, in terms of services offered, terrestrial analogue televisions typically consist of 5 main channels (BBC1, BBC2, ITV, Channel 4, and Channel 5). Apart from the television programs themselves, the only additional functionality commonly provided is teletext. This is a means by which pages of text-based information are displayed on the screen. Typical teletext pages include news summaries, weather forecasts, and travel information. In many respects the service is quite limited, being restricted to displaying 24 rows of up to 40 characters per row and capable of only displaying 7 different colors. However, the teletext system has proved remarkably popular with viewers.

By keeping the level of detail low, then each television channel could offer (in theory) several thousand pages of text piggybacking on the main program signal. These pages are typically accessed through pressing a dedicated TELETEXT button on the remote control and then through page number navigation. This involves either typing the desired page number (a three digit number) or else using the color-coded Fastext shortcuts (the red, blue, green, yellow buttons on the remote control), the equivalent of a hypertext shortcut link on a web page.

Perhaps the most important service offered through teletext is subtitles—a version of closed captioning. Subtitles are always found on page 888, no matter which channel is being viewed. A study in the UK showed that of the 7.5 million viewers who use subtitles, 6 million have no hearing impairment. In other words, a service that was primarily intended to benefit 1.5 million people with hearing difficulties also benefits approximately *four times* that number with no such difficulty (Duffy, 2006). There are many reasons why this service is so popular, for example parents watching programs while trying to get a baby to sleep, others keeping track of a program while someone else is using the telephone, people who have difficulty interpreting the different regional accents or actors' poor pronunciation can follow what is being said. Subtitles (and closed captioning) represent an excellent example of how accommodations intended for one particular user group can often also benefit other users.

Digital television offers many more channels (more than 50 free-to-view channels), along with radio broadcasts. There are channels that are dedicated to providing digital text (the digital replacement for teletext) only, rather than teletext being available as an add-on to the 5 main channels on analogue. The direct teletext replacement offers interactive content, such as for BBCi and is usually accessed through pressing the red "Fastext" button on the remote control. The final significant difference is the presence of numerous on-screen menus, ranging from electronic program guides (the EPG) to language and subtitle settings.

In summary, digital television offers significantly more functionality than analogue television. However, the corollary of that is that the interaction is potentially more complex and thus may exclude more possible users.

FRAMEWORK FOR UNDERSTANDING THE INTERACTION

To understand the interaction, it is necessary to have a framework for describing and interpreting the processes being observed. In this case, it was appropriate to use a human-computer interaction (HCI) type model as the basis for this study as this allows a focus on the key elements of user interaction. For example, the Model Human Processor approach identifies perception, cognition, and motor functions as the building blocks for any interaction (Card, Moran, and Newell, 1983). These may be further decomposed into basic elements as discussed in Chapter 10:

- **perception (sensory)**—vision and hearing;
- **cognition**—communication and intellectual functioning;
- **motor**—locomotion, reach and stretch, and dexterity.

The advantage of this approach was that it also facilitated quantitative and qualitative investigation. The basic elements of interaction could be observed and data would become available for the identification of the prevalence of such characteristics in the general population.

Older Adults and Digital Television

Older adults already make up a significant proportion of the population, and that proportion is set to continue increasing in most developed countries. They are also becoming increasingly politically active as a group (GAD, 2001), and therefore the UK Government was especially eager to ensure that older adults, as a whole, were not excluded by the move to digital television. There are two potential causes of exclusion of older adults when considering the use of STBs:

1. Physical accessibility—e.g., can they read the on-screen display and press the buttons on the remote control?
2. Mental model of the interaction—e.g., can they understand what the system is doing and successfully predict the outcome of any action they perform?

Physical Accessibility

As Table 12–1 shows, as people get older, the prevalence of functional impairments as a percentage of the population increases (Hirsch et al., 2000). Reduced levels of capability can arise from many causes. These include medical conditions such as strokes and Parkinson's disease, trauma (accidents), and the process of ageing. No attempt is made in this study to distinguish between any of these causes, as it is the effects of the symptoms, i.e., the reduced capabilities of the user, that are of principal importance when interacting with the STBs. The levels of data shown in Table 12–1 are calculated from the 1997 UK Disability Follow-up Survey (Grundy et al., 1999) to the 1996/7 Family Resources Survey (Semmence et al., 1998).

TABLE 12–1
Numbers of People in Great Britain With Some Loss of Capability
for Ages 16–49 and 75+ (Grundy et al., 1999)

Age Bands:	16–49		75+	
	,000s	%	,000s	%
Motor	1,484	5.4	1,970	47.2
Sensory	617	2.3	1,603	38.4
Cognitive	862	3.2	615	14.7
Total	1,975	7.2	2,442	58.5

Mental Model of Interaction

It is widely accepted that for successful interaction, the user requires a clear and unambiguous mental model of the interaction. The interaction paradigms for analogue television have remained comparatively unchanged since the technology was introduced in the 1930s. The only real changes have been the advent of pushbutton controls instead of tuning dials, remote controls and teletext.

Digital television, however, introduces completely new concepts to many viewers. Instead of changing channels through the television set, the set-top box becomes the focus of the interaction. Many of the interactive features have been designed following computer input paradigm, such as the need to press an OK / SELECT button to activate an on-screen option and the use of arrow keys to navigate through menus. These are potentially new paradigms for viewers, especially older adults who may not be familiar with computer technology.

Trying to identify whether most difficulties with accessibility were associated with physical accessibility or the user's mental model is an interesting topic. Many accessibility assessments focus almost exclusively on physical accessibility (e.g., DTI, 2000), because physical impairments are easier to recognize, simulate, understand, and assess. One of the aims of this study was to identify the balance between physical accessibility difficulties and those arising from the discrepancy between the system and the user's mental model.

THE STUDY

The number of STBs is increasing rapidly. For the purposes of this study, efforts were focused on looking at two set-top boxes selected to be representative of the systems currently available. These will be referred to throughout the chapter as STB1 and STB2. STB1 was selected because it was being marketed as "easy to use" and STB2 was chosen because it was the market leader at the time.

In assessing the accessibility of STBs, it is important to understand the contribution of each of the elements to the potential for exclusion. The system must therefore be tested as a whole and in a way that represents "normal" use. It is also important to remember that "use" starts with the purchasing and commissioning of the system. As a result, a number of "use-case scenarios" were used to investigate the accessibility of DTV focusing on the purchasing, installation and use of STBs. These are shown in Table 12–2.

The investigation focused on identifying the broad steps involved in the interaction between the user and the STBs. The aim was to identify the potential causes of exclusion that may prevent users from interacting with the STBs effectively. Typical approaches to such accessibility studies include expert assessment (using systematic analyses), simulation, and user trials (Keates and Clarkson, 2003; Popovic, 1999). In this case, it was decided to focus on user observations following a pilot expert assessment study.

Expert Assessment

The purpose of the pilot expert assessment was to provide an indication of where users may have difficulty interacting with a typical DTV system. Four users participated in mock user observation sessions. Two were experienced usability practitioners with extensive DTV knowledge and the other two were experienced usability engineers, but digital television novices.

In terms of summarizing what was learned, the following summary points provide a list of key challenges encountered when assessing the STBs. Particular atten-

TABLE 12–2
The Use-Case Scenarios for the Digital Television Study

Scenario	Title	Description
1	System purchasing	How easy is it to make an informed choice which STB to buy in a shop or from a Web site? (e.g., how to tell the difference between them; which one is easier to use? etc.)
2	System set-up requirements	How easy is it to identify the set-up requirements? (e.g., what cables do I need?)
3	Installation instructions	How easy are the instructions to read (physical accessibility—size of print, etc.) and to understand?
4	System installation	How easy is the STB to connect to the television?
5	System tuning	How easy is the STB to tune in on first use?
6	Channel selection	How easy is it to find out what programs are available and select the preferred channel (using either the interactive electronic program guide (EPG), or by random surfing)?
7	Subtitles	How easy is it to find and operate the subtitles function?
8	Teletext operation	How easy is it to find and navigate a particular teletext page?
9	Interactive content operation	How easy is it to find and navigate the comparable page using the interactive digital content?

tion has been given to identifying general problems that limit their accessibility, rather than problems specific to an individual box. Following analysis of the expert assessment, a list of points of comment or concern was prepared.

Installation and Set-Up

1. **Set-up requirements**—Retailers appear to have inadequate knowledge of installation requirements (signal coverage/strength, need for aerial, etc.) and the necessary additional equipment is not always available at the point of purchase of the STB. There is also then an additional cost required to complete installation. This is not a problem for equipment which is "installed" for the user as part of the purchased package (e.g., with STB3).

2. **Set-up instructions**—These are critically important for the users who attempt to self-install their STB. The instructions need to be clear and up to date (e.g., not referring to an earlier model of remote control as was the case with STB2). Graphical instructions, if done well, can be particularly helpful if they identify buttons on the remote control and elements on the display that the user has to interact with. A complementary combination of graphical and written instructions minimizes the potential for exclusion.

3. **Tuning**—The initiation of the tuning of the STB must be straightforward, both in terms of when it must be done and how it must be done. Clear instructions must be provided on both these points.

4. **Battery replacement**—Remote controls use batteries, therefore it is important that access to the battery compartment is straightforward without compromising the safety of small children. The need to reactivate the remote after replacing batteries should also be avoided since this is unnecessary and uncommon practice.

Operation

1. **Operating modes**—There is potential for confusion regarding STB and TV modes and the buttons required to switch between them. Simpler operation and clearer instructions are required.

2. **Volume controls**—The presence of direct TV volume controls on the remote control is ambiguous. On only one of the remotes used did this feature actually work. In practice, this would be a very useful feature, allowing channel selection and volume adjustment from the same remote control.

3. **Response times**—The delay between pressing buttons on the remote and seeing a response on the TV screen is unacceptably long and is likely to lead to much confusion. Unless the user is familiar with the system and has the confidence that it will respond, many actions are interrupted by the user because no feedback has been provided that the user request has been received. Good practice is to ensure that the user is provided with direct and immediate feedback in response to an action (e.g., by using an hourglass symbol).

4. **On-screen menus**—The use of on-screen menus must be done in such a way as to be intuitive for inexperienced users. The selection of highlighted options provides a particular cause of confusion since it is not immediately clear if the default selected item (usually at the top of the list) is a heading or a member of the list of options. This is particularly ambiguous for novice users and those unfamiliar with menu-driven systems.

5. **Subtitles**—The accessibility of subtitles is important. Expecting users to navigate multiple menu layers with ambiguous names (e.g., "language") is not acceptable. A single button press is better.

Remote Controls

1. **Nomenclature**—There is a lack of consistency in the use of nomenclature by the various providers of digital television. This, coupled to the use of different on-screen navigation paradigms, leads to unnecessary confusion. Some menu systems require navigation using cursor keys and a SELECT/OK button, others rely on selection by number. The latter is preferable for those without experience of menu-driven systems.
2. **Labeling**—Appropriate fonts of appropriate sizes in appropriate colors should be used for labeling on the remote control.
3. **Layout**—Care needs to be taken that commonly used keys on the remote are not so close together as to invite the user to inadvertently select the wrong function. In addition, the legends used need to be informative to those without general experience of such devices. Confusion can arise between cursor control keys that are often unmarked and the keys used for program selection that are often marked with "up" and "down" arrows.
4. **Sourcing**—The use of off-the-shelf remote, generic controls (rather than ones designed specifically for the particular function required to operate an STB) can provide buttons that are not used and layouts that are not suited to the system they should control. A remote designed in conjunction with the system it controls is likely to be better suited to the task.

This pilot study provided a valuable opportunity to refine the use-case scenarios shown in Table 12–2, as well as provide a basis for estimating how long each scenario would take. It was important that each user trial session lasted less than two hours, with the operation of the equipment phases lasting less than one hour of each session to avoid the effects of fatigue affecting the data collected.

User Trials

Building on the results of the earlier assessments, a series of user observation sessions were conducted. As we have examined throughout this book, user trials are an invaluable tool when assessing both the usability and accessibility of a product.

User Selection

A total of thirteen users were recruited and the observation took place over a period of six days. The users were identified based on a number of criteria, primarily focused around whether they were strong candidates for being edge-cases in terms of their ability to interact with STBs. It was decided to focus on recruiting older adults not living in residential care. More extreme users could have been selected for the user group, however, the level of information that can be obtained is then limited.

The decision to select users still living in private homes, rather than residential care was based on the desire to have users who still have enough functional capability to support independent living to some degree. They should therefore be able to perform tasks such as operating a television on their own. If they experienced significant difficulty, then it could be argued that the STBs are causing undue exclusion. As a comparison, a younger user with a more severe impairment was also recruited to highlight whether a more severely impaired user would also encounter the difficulties experienced by the older adults.

During the recruitment process efforts were made to ensure that the users exhibited a range of capabilities. For example, it was known that those with arthritis would exhibit a loss of dexterity, that many older users would exhibit macular degeneration or loss of hearing, and that it would be very likely that a range of intellectual functioning capability would be observed. Care was taken not to skew the sample towards any particular capability loss, rather to provide a balanced representation of motor, sensory and cognitive losses.

The Users

To maintain consistency with the earlier STB assessments the users were asked to self-report their own assessments of their capabilities, from which their scale points were deduced. This is the same approach as the ONS adopted in their data collection.

- User 1 was a retired academic. He had limited locomotion capability, and walked with frequent, short steps, using a walking stick for support. This user also had a slight, but constant tremor in his hands, making activities involving hand-eye co-ordination difficult. He also wore spectacles and a hearing aid on one ear.

- User 2 was a retired civil servant. She self-reported only a mild hearing impairment, but had difficulty following many of the instructions during the course of the observation session. She also exhibited a tendency to forget things and occasionally lost track during conversations. Otherwise she exhibited no signs of capability loss, beyond being slightly slower in her physical movements than would be expected for a younger person.

- User 3 had a range of minor impairments, covering all of the motor, sensory, and cognitive capabilities. However, none of the impairments appeared to have a specific medical cause and may be attributed to the ageing process. This user led an active life, including cooking lunch for other members of her day-care group. She appeared to be nervous around high technology. She had tried to learn to use a computer in the past, but gave up on it because, in her own words, her "memory isn't very good."

- User 4 was a retired nurse. She was a wheelchair user, and was thus unable to participate in any of the activities involving locomotion. Her only other noticeable impairment was a need to wear reading glasses to read small print (such as button legends on a remote control). She regarded many high-technology products with disdain, proudly declaring her house to be "a computer-free zone."

- User 5 was a retired member of the clergy, who exhibited a strong affinity for high-technology products. He was one of the few participants who would have bought an STB with a view to installing it himself. This user exhibited mild capability loss in all three categories, i.e., sensory, motor, and cognitive. As with many of the other users, there was no apparent medical cause for these losses and so were attributed to the aging process.

- User 6 was a retired deputy-head teacher, who only exhibited a mild loss of cognitive capability. She also displayed a fear of high-technology products, often relying on her husband or children to operate the STB that she had at home.

- User 7 was also a retired deputy-head teacher, who was a self-professed fan of high-technology products. He did not appear to exhibit any capability loss, beyond a mild hearing impairment.

- User 8, despite being a computer manager during his working life, does not consider himself to be very adept with high-technology products. However, he showed himself to be willing to try to learn new things, even though he did not rate his chances of success as terribly high. He exhibited no obvious capability loss.

- User 9 was a wheelchair user who exhibited moderate to severe loss of motor capability, including the inability to use one hand through arthritis. This user needed to use distance glasses to read the television on-screen text, but reading glasses to read the button legends on the remote control. Thus, she had to keep changing between two remote controls and two pairs of glasses, while having the use of only one hand. Although she did not self-report any, she appeared to exhibit a degree of loss of cognitive capability as well and had difficulty understanding sections of the user observation sessions.

- User 10 had no discernible impairment beyond having an artificial eye, although she had a number of severe illnesses in the year or so before participating in these user observation sessions. She was very dismissive of high technology and frequently repeated that she was too old to be learning new things and that she probably "would not be around" when analogue television was switched off.

- User 11 exhibited a range of mild sensory and moderate motor capability losses. However, she was a very active individual and along with user 3, frequently prepared lunch for the other members of the day-care group.

- User 12 was a wheelchair user with moderate cognitive and mild vision capability loss. However, on the day of the user observation session, she forgot to bring her distance glasses, and so had to use her reading glasses throughout. This resulted in difficulty reading the on-screen text, and so all such text had to be read aloud to her.

- User 13 was the youngest participant in the trials by approximately 35 years. He exhibited the most severe sensory loss, being registered as having a severe and permanent loss of sight, and being eligible for legal classification as blind. His particular visual condition was congenital, and primarily involved loss of central field of view. Otherwise he was fully able. He was also a PhD student in Computer Science and thus had a strong affinity for high-technology products.

User 13 was involved in the trials as a counterpoint to the other users. His visual impairment was more severe than for any of the other users; however he was highly experienced with computers and their interaction paradigms. Thus, it was anticipated that he would encounter primarily physical accessibility difficulties, rather than those associated with the disconnect between the user's mental model of the interaction and the system's design.

It is worth noting that a range of user capabilities was observed. Two users showed no obvious impairment, four showed single impairments, and the remaining seven exhibited multiple impairments. Three of the users reported a loss of dexterity at levels likely to cause difficulties using the STBs. Five of the users reported a loss of intellectual functioning, while three were at levels that might also be likely to cause difficulties.

A number of the users were computer-literate and familiar with digital television services. They were selected to provide an insight into the effect of prior experience of digital television services and computer-based menu systems on the use of otherwise unfamiliar STBs.

Methodology

The user observation sessions were organized to be limited to 2 hours to ensure that user fatigue was kept to a minimum. Thus the user activities were restricted to those operations that could be considered fundamental to watching television, such as the ability to change channels, and also to those advanced features that could be explored within the available timeframe. Two STBs were assessed (STB1 and STB2), to ensure a balance between breadth and depth of study. The STBs selected for the trial reflected different design approaches, with one focused more on ease-of-use and the other on functionality.

Initially the users were interviewed for 30 minutes to find out their capability profiles and also background information on their attitudes towards television use and exposure to DTV. Two or three observers attended each interview, each recording the user responses. Following the interview, the users began an equipment trial. This commenced with a familiarization exercise with the analogue television set being used. All users used the same television and remote control. The users were asked to perform basic operations, such as changing channel and volume. They were also asked to use teletext services and to call up subtitles.

The users were then asked to choose which of the two STBs being assessed they would prefer to buy. This involved showing them the external packaging and then the STBs themselves [Scenario 1—System purchasing]. The users were then asked questions about whether the STB would work with the television set-up provided, such as whether the television set had the right connections to hook up or if the aerial was rated for receiving digital signals [Scenario 2—System set-up requirements].

The next stage was to provide the users with the installation instructions for their chosen STB [Scenario 3—Installation instructions] and to ask whether they would install the box themselves. Those users who felt up to doing so were encouraged to connect up the STB to the television [Scenario 4—System installation]. For those users who declined to do so, the STB was connected for them. For the final set-up task, the users were asked to switch on the STB and complete the tuning procedure [Scenario 5—System tuning].

The subsequent stage was to look at the operation of the STB, particularly simple, everyday television operations such as changing channels and channel hopping. Users were encouraged to use the on-screen electronic program guide (EPG) for one of the channel hops [Scenario 6—Channel selection]. The more advanced interaction activities included finding weather and television program guide information from both Teletext [Scenario 8—Teletext operation] and BBCi (the BBC's new interactive digital service) [*Scenario 9—Interactive content operation*], as well as calling up subtitles [*Scenario 7—Subtitles*].

The equipment trial took an average of one hour to complete. Finally, a closing de-brief session was held, that lasted approximately 15 minutes. During this session, the users were asked what they thought of their experience with the STBs.

Results

Throughout the assessment on accessibility of DTV set-top boxes, interaction was considered in terms of the sensory, cognitive, and motor demands placed on the users. Tables 12–3 and 12–4 provide a summary of the incidence of difficulties experienced.

Note that Tables 12–3 and 12–4 represent a very concise summary of an extensive data collection activity. For a full description of the data collected, including the raw data from the observation sessions and detailed calculations of the number of users excluded, please refer to the full DTI report (DTI, 2003).

As can be seen from Table 12–3, there were 10 unique difficulties encountered across all 13 users when interacting with the analogue television. However, Table 12–4 shows that there were 65 unique difficulties for the STBs. It is also notable that for 4 out of the 12 activities involving the STB, all 13 users experienced usability difficulties. This appeared to be independent of the functional capabilities of the users. Note that a usability difficulty is one requiring prompting from the observer to overcome or rectify the difficulty.

Common sensory problems included finding/reading buttons on the remote controls, reading on-screen text, and swapping between the two (especially for users with distance and reading glasses). These problems are made worse in comparison to analogue television because of the increased functionality leading to the need for more (and hence smaller) buttons and also increased use of on-screen text displays. Users with hearing impairments would find the presence of an explicit subtitle button on the remote control for STB1 very useful, but would be disadvantaged by the

TABLE 12–3
The Distribution of Causes of Difficulty for Analogue Televisions for the 13 Users

| Activity | # of Users Having Difficulty | Problem Classification | | | # of Unique Difficulties |
		Motor	Sensory	Cognitive	
Switching on	8	1	—	1	2
Changing to a specified channel	1	1	—	—	1
Channel-hopping	—	—	—	—	—
Changing volume	—	—	—	—	—
Using teletext	6	1	3	3	7
Using subtitles	—	—	—	—	—

TABLE 12–4
The Distribution of Causes of Difficulty for Digital Television Services for the 13 Users

Activity	# of Users Having Difficulty	Problem Classification			# of Unique Difficulties
		Motor	Sensory	Cognitive	
Connecting the STBs	4 (out of 6 users)	1	—	1	2
Switching on the television	1	1	—	—	1
Switching on the STB	6	—	1	2	3
Changing DTV channels	3	—	1	2	3
Changing volume	7	—	—	1	1
Changing to a high channel number	10	2	2	2	6
Changing channel via the EPG	13	3	4	6	13
Teletext	13	2	4	12	18
Subtitles—button	6	—	1	1	2
Subtitles—menu	13	—	1	4	5
BBCi	13	2	1	6	9
Switching off	5	—	—	2	2

on-screen menu approach of STB2, where the user had to navigate through several levels of menu to reach the subtitles option.

The most common source of motor difficulties was pressing the buttons on the remote control. Again, while this is a common task for both analogue television and the STBs, it is made more difficult for the latter by the need for more (and hence smaller) buttons and also increased levels of user interaction.

However, while there was an increase in both the vision and dexterity demands made upon the users, by far the biggest cause of exclusion noted during the user observation sessions was the cognitive demands. The inherent increase in user cognitive effort associated with having to use two remote controls (or a single remote control with multiple modes) rather than a single remote control is further exacerbated by the mismatch between the users' mental models of the interaction and the interaction paradigms adopted. For example, users are familiar with the concept that pressing a button on a television remote control has an immediate effect on what they see on the screen. For example, pressing a channel number button causes the television to tune immediately to that channel. Thus a strong link between cause and effect is observed, and a solid user mental model of the interaction is developed.

The STBs, though, present the users with numerous new interaction paradigms, such as pop-up menus, combined with weakened cause and effect. For example, nothing happens when an item is highlighted on a pop-up menu until the OK/SELECT button is pressed (another new concept). The situation is worsened further by the seemingly arbitrary inconsistencies in language and interaction between similar purpose entities of the interface. For example, in BBCi the menu option is called menu, whereas in Teletext it is control. On one remote control the SELECT button was called just that, whereas on the other it was denoted OK. To enter BBCi, the user has to press the RED button, while for Teletext it is the TEXT button. These inconsistencies present unnecessary usability hurdles to the users. These differences breach one of the central tenets of usability theory, namely that of the need for consistency.

CONCLUSIONS

This user study has highlighted a number of important issues when designing for older adults.

Mental Model vs. Physical Accessibility

The majority of difficulties encountered were cognitive in origin, unlike for the analogue television, where the sources of difficulties were evenly distributed. Table 12–5 shows a summary of the distribution of causes of difficulty across all 13 users.

Many of the cognitive difficulties experienced were not directly attributable to any kind of "medical model" impairment. Instead, lack of experience with, and mental model of, the interaction paradigms used in digital television was the principal cause of the difficulties encountered. This is an important finding for research into assessing the accessibility (or inclusivity) of systems, because it means that estimates of the numbers of people excluded from using particular systems or products based purely on physical accessibility assessments are almost certainly going to be conservative. This is particularly true if no allowance is made of cognitive issues, during the assessment.

Of particular importance when considering IT systems is that of (lack of) past experience for the users, especially when considering older adults. This is a recurring problem, having also been seen in a study I performed for the Royal Mail, looking at kiosk design for older adults. In the case of the kiosk assessment, the designers had chosen inappropriate icons to represent certain functions of the kiosk. For example, the icon for obtaining a paper printout looked like a standard desktop printer, as would be expected on a standard graphical user interface (GUI). However, the

TABLE 12–5
The Distribution of Causes of Difficulty

	Vision	Dexterity	Cognitive
Analogue (out of 10)	3/10 (30%)	3/10 (30%)	4/10 (40%)
STBs (out of 65)	15/65 (23%)	11/65 (17%)	39/65 (60%)

users (all over the age of 70) had never seen a desktop printer and could not make the necessary cognitive connection.

These difficulties often arise, because the designer has assumed that the user's mental model of the interaction is the same as his/her own. This is the back to the "designers design for themselves" phenomenon (Cooper, 1999). However, given the very different background experiences and expertise between (typically young) designers and older adults, it should come as no surprise that there are differences between their mental models of interaction. Thus it is important that products and systems that are designed to be used by older adults are subject to appropriate accessibility assessments.

While there are methods of performing physical accessibility assessments in the absence of users, for example through simulation and the use of user models, there does not appear to be a viable or feasible substitute for user observation sessions when considering mental model difficulties. This result has strong implications for companies designing products and systems for older adults.

Suggested Areas for Re-design

For the case of STBs, a number of key areas for re-design can be identified. Several of these are shown in Table 12–6. This is not a complete list; it is intended to highlight the origins of some of the common problems and how they may be countered.

Having identified the most prominent problems encountered by the users, the next step is to propose potential solutions in a format that designers can readily use and interpret. Tables 12–7 and 12–8 show examples of how information about the problems and their potential solutions could be presented to designers. The full report into this study details all of the design recommendations made to the Department of Trade and Industry (DTI, 2003).

A study of the possible fixes (e.g., those in Tables 12–7 and 12–8) show that many are complementary and that some solutions may benefit more than one problem. For example, the use of appropriate affordances (Gibson, 1977) could aid the user's

TABLE 12–6
Key Areas for Re-design

Category	Problem
High-level understanding:	1: Poor user mental model of interaction
	2: Inconsistent language/labeling
Performing interaction:	3: Poor accessibility of remote control
	4: Multiple modes
	5: Use of OK/SELECT button (on remote control)
	6: STB times out on user input
	7: Delay in responding to user input
	8: Switching on subtitles via menus
Instructions:	9: Unclear set-up instructions
	10: Unclear instructions for use

TABLE 12–7
Problem 6: STB Times Out on User Input

Symptoms:
- STB only responds to part of the input from the user.

Encountered:
- When changing channel, ending up on channel 04 instead of channel 40.

Result:

Channel changing
- Impairment of ability to change channels.
- Tendency to end up on wrong channel.
- Reinforces concept of digital television "being difficult."

Causes:

Principal cause
- Time-outs almost certainly based on model of young, able-bodied user.

Exacerbated by
- Checking the screen to see that the previous input has been recognized (e.g., showing "4-" when trying to go to channel 40)—even further exacerbated if this involves a change in spectacles from reading to distance pairs.
- Finding the next button to press (e.g., the "0" on the STB1 remote is not in the standard position).
- Deciding on the next button to press (e.g., the OK/SELECT button).

Possible fixes:
- Extend time-out periods to allow for "slower" users.
- Use of warning-style dialogue boxes before dropping a user out back at the start of a process ("You have not selected an option—do you wish to do so or leave the menu?"—then press appropriate button)—Problem: This may get annoying if you keep encountering it.
- Use of buttons such as the "-/--" button on remote controls for specifying "I am doing a two-digit input now"—Problem: Not many users understand what this button does.
- Reducing the "exacerbating" features—e.g., having an LCD display on the remote control show the user the input created (saves having to swap between the TV and the remote control)—ensuring "standardized" layout of remote controls to reduce hunting for buttons—fixing the OK/SELECT problem.

---

mental model of the interaction, at the same time as providing reminders to use the OK/SELECT button.

The Future of Digital Terrestrial Television

STBs will only cease to exclude more people than analogue televisions when their operation is completely transparent from the user's point of view. Integrated digital televisions, for example, appear to manage to achieve this level of transparency for basic functions by using only a single remote control with minimal need for mode changes.

TABLE 12–8
Problem 7: Delay in Responding to User Input

Symptoms:
User presses a button and either nothing seems to happen or the screen goes blank.

Encountered:
- Changing channel (initial black screen from inherent digital television lag on channel change, no indication of which channel moved to).
- Calling up interactive elements (e.g., BBCi).
- Calling up subtitles.

Result:
- User believing that certain functionality is not available.
- User repeatedly pressing button (potential for ending up somewhere unexpected, damaging user mental model of interaction).
- User frustration.

Causes:
- Download time for system to update screen.

Possible fixes:
- Clear, unambiguous feedback from the STB that the input has been recognized—e.g., teletext page "loading" legend.

However, a subsequent extension to the study presented in this chapter showed that even iDTVs exclude more people than analogue televisions when considering the full range of operation.

Put simply, digital television offers more functionality, and thus requires more cognitive effort to learn and operate. For example, if a user wishes to use the full functionality of DTV, then there is a greater need to be able to read the on-screen display and to swap to reading the remote control (vision demand). Similarly, the users need to be able to operate the arrow buttons and SELECT/OK, rather than just the channel numbers. The increase in the number of channels means that users have to enter more double-figure channel numbers, with the inherent time-out limitations increasing the dexterity demand still further.

Only if all of the additional functionality is as accessible and usable as interacting with an analogue television, will digital television not excluding any one who can currently use an analogue television. This is a tough target to aim for, but a necessary one unless it is to be accepted that not all users will have access to all of the digital services. It would take a brave government to switch off a broadcast service that people are currently able to access and so, by which they are enabled to watch television, and replace it with one that they could not access.

The predominance of exclusion arising from the differences between the users' mental models and the interaction paradigms within the interface affects far more users than those that would typically be classed as a stereotypical "special needs user". This is well illustrated by the comparative lack of difficulty with the interaction experienced by the youngest participant who had the most severe vision impairment of any of the users, but who, nonetheless, experienced little difficulty completing the tasks, most probably because of his wide experience with high-technology products.

Consequently, manufacturers should be encouraged to look beyond the stereotypes of young, severely impaired people when considering who may have difficulty using their STBs and also to consider the needs of older adults and those who may not be familiar with the interaction paradigms used. There is also a clear need to standardize within those paradigms to minimize the cognitive demand placed on the users and to make interaction with the STBs as transparent as possible.

Ultimately, what is being advocated is not special purpose design for a small market sector, but rather good "design for all."

And that, dear reader, seems like a very apt point on which to finish.

KEY POINTS

- Constructive suggestions on how to remedy difficulties are better than just criticizing a product's failings.
- Identifying the cause and nature of difficulties using a product often provides pointers to how to fix them.
- Exclusion does not arise solely from physical limitations—a poorly constructed mental model can exclude just as much.

References

ACM. (2005). *The ACM digital library*. Retrieved January 14, 2006, from http:// portal.acm.org/dl.cfm

ACSO. (2006). *American Community Survey 2004 subject definitions*. American Community Survey Office, U.S. Census Bureau. Retrieved March 15, 2006, from http:// www.census.gov/acs/www/Downloads/2004/usedata/Subject_Definitions.pdf

ADA. (1990). *Americans with Disabilities Act* (U.S. Pub. L. 101-336). Retrieved January 14, 2006, from http://www.usdoj.gov/crt/ada/pubs/ada.txt

Akao, Y. (1990). An introduction to quality function deployment. In Y. Akao (Ed.), *Quality Function Deployment: Integrating customer requirements into product design* (pp. 1–24). Portland, OR: Productivity Press Inc.

Aus DDA. (1992). *Disability discrimination act 1992* (Australia). Retrieved January 13, 2006, from http://unpan1.un.org/intradoc/groups/public/documents/apcity/unpan 004021.pdf

Beitz, W., and Kuttner, K.-H. (1994). *Handbook of mechanical engineering*. London: Springer-Verlag.

Benktzon, M. (1993). Designing for our future selves: the Swedish experience. *Applied Ergonomics, 24*(1), 19–27.

Blessing, L. T. M., Chakrabati, A., and Wallace, K. (1995). A design research methodology. In *Proceedings of International Conference on Engineering Design 1995* (Vol. 1, pp. 50–55). Prague, Czech Republic.

BSI. (1997). *BS 4467:1997—Guide to dimensions in designing for elderly people*. London: British Standards Institution.

BSI. (1999). *BS7000-1:1999 Design management systems. Guide to managing innovation*. London: British Standards Institution.

BSI. (2001). *BS 8300:2001 Design of buildings and their approaches to meet the needs of disabled people. Code of practice*. London: British Standards Institution.

BSI. (2005). *BS 7000-6:2005 Design management systems. Managing inclusive design. Guide*. London: British Standards Institution.

Buhler, C. (1998). Robotics for rehabilitation—A European(?) perspective. *Robotica, 16*(5), 487–490.

Card, S. K., Moran, T. P., and Newell, A. (1983). *The psychology of human-computer interaction*. Hillsdale, NJ: Lawrence Erlbaum Associates.

Cassim, J. (2004). Cross-market product and service innovation—the DBA Challenge example. In Keates et al. (Eds.), *Designing a more inclusive world* (pp. 11–19). London: Springer-Verlag.

CDF. (2001). Getting a grip on kitchen tools. @issue, *Corporate Design Foundation*, (2)1.

Chin, D. N. (2000). *Empirical evaluation of user models and user-adapted systems.* Department of information and Computer Sciences, University of Hawaii, 1680 East West Rd., Honolulu, HI 96822, USA.

CHRA. (1985). *Canadian human rights act* (R.S. 1985, c. H–6). Retrieved January 15, 2006, from http://laws.justice.gc.ca/en/H–6/

Cleveland T. (2002). Accessibility laws in Canada. *evolt.org*. Retrieved January 14, 2006, from http://www.evolt.org/article/Accessibility_Laws_In_Canada/4090/28074/

Cohen, J. (1988). *Statistical power analysis for the social sciences.* (2nd ed.). Hillsdale, NJ: Lawrence Erlbaum Associates.

Coleman, R. (1993). A demographic overview of the ageing of first world populations. *Applied Ergonomics, 24*(1), 5–8.

Coleman, R. (2005). About: Inclusive Design. *Design Council*. Retrieved January 12, 2006, from http://www.designcouncil.org.uk/inclusivedesign/

Cooper, A. (1999). *The inmates are running the asylum.* Indianapolis: SAMS Publishing.

Coy, J. (2002). Commercial perspectives on universal access and assistive technology. In Keates et al. (eds.), *Universal access and assistive technology* (pp. 3–10). London: Springer-Verlag.

Dahlgard J. J., Kristensen, K., and Khanji, G. K. (2005) *Fundamentals of total quality management.* New York: Routledge.

DDA. (1995). Disability discrimination act 1995 (c. 50). Retrieved January 11, 2006, from http://www.opsi.gov.uk/acts/acts1995/Ukpga_19950050_en_1.htm

DfA:AT. (2005). Design for all and assistive technology (DfA & AT) Awards. Retrieved January 12, 2006, from http://www.dfa-at-awards.org/home/index.cfm

Dowland, R., Clarkson, P. J., and Cipolla, R. (1998). A prototyping strategy for use in interactive robotic systems development. *Robotica. 16*(5), 517–521.

DRC. (2005). Codes of practice: How they help. *Disability Rights Commission.* Retrieved January, 12, 2006, http://www.drc-gb.org/thelaw/practice.asp

DTI. (2000). *A study on the difficulties disabled people have when using everyday consumer products.* London: Government Consumer Safety Research, Department of Trade and Industry.

DTI. (2003). *Digital television for all—a report on usability and accessible design.* London: Department of Trade and Industry. Retrieved January 14, 2006, from http://www.digitaltelevision.gov.uk/publications/pub_dtv_for_all.html

DTI Foresight. (2000). *Making the future work for you.* London: Department of Trade and Industry.

Duffy, J. (2006). The joys of subtitles. *BBC News Magazine.* Retrieved March 31, 2006, from http://news.bbc.co.uk/2/hi/uk_news/magazine/4862652.stm

Europa. (2000). Charter of fundamental rights of the European Union (2000/C 364/01). *Official Journal of the European Communities, European Commission.* Retrieved January 14, 2006, from http://www.europarl.eu.int/charter/pdf/text_en.pdf

Europa. (2002). Consolidated version of the treaty establishing the European Community. Retrieved January 15, 2006, from http://europa.eu.int/eur-lex/en/treaties/dat/C_2002 325EN.003301.html

Europa. (2005). Design for Retrieved January 14, 2006, from http://europa.eu.int/information_society/policy/accessibility/dfa/index_en.htm

Fitts, P. M. (1954). The information capacity of the human motor system in controlling the amplitude of movement. *Journal of Experimental Psychology, 47*(6), 381–391.

Follette Story, M. (2001). The principles of universal design. In Preiser, W., and Ostroff, E. (Eds.), *Universal design handbook*. New York: McGraw-Hill.

GAD. (2001). *National population projections: 2000-based*. London: Office for National Statistics.

Gheerawo, R. R., and Lebbon, C. (2002). Inclusive design—developing theory through practice. In Keates et al. (Eds.), *Universal access and assistive technology* (pp. 43–52). London: Springer-Verlag.

Gibson, J. J. (1977). The theory of affordances. In Shaw and Bransford (Eds.), Perceiving, acting, and knowing. Hillsdale, NJ: Lawrence Erlbaum Associates.

Goyder, M. (2005). Pensions crisis? What crisis? *BBC Online*. Retrieved January 13, 2006, from http://news.bbc.co.uk/2/hi/business/4470088.stm

Grundy, E., Ahlburg, D., Ali, M., Breeze, E., and Sloggett, A. (1999). *Disability in Great Britain: Results from the 1996/7 disability follow-up to the Family Resources Survey*. Huddersfield, UK: Charlesworth Group.

Hirsch, T., Forlizzi, J., Hyder, E., Goetz, J., Stroback, J., and Kurtz, C. (2000). The elder project: social, emotional and environmental factors in the design of eldercare technologies. In *Proceedings of the 1st International Conference on Universal Usability* (pp. 72–79). New York: ACM Press.

Hurst, M. (2005). *This is broken*. Retrieved January 14, 2006, from http://www.thisisbroken.com/

IBM. (2005). *aDesigner*. Retrieved January 15, 2006, from http://www.alphaworks.ibm.com/tech/adesigner

IBM. (2005a). *Human ability and accessibility center*. Retrieved January 14, 2006, from http://www.ibm.com/able/index.html

IBM. (2005b). *Developer guidelines*. Retrieved January 14, 2006, from http://www.ibm.com/able/guidelines/index.html

IBM. (2005c). *Conducting user evaluations with people with disabilities*. Retrieved March 14, 2006, from http://www.ibm.com/able/resources/userevaluations.html

ISO. (2000). Quality management systems. *ISO 9000:2000 series*. Geneva: International Organization for Standards.

ISO. (2001). *PD ISO/IEC Guide 71—Guidelines for standards developers to address the needs of older persons and persons with disabilities*. Geneva: International Organization for Standards.

ISO. (2002). *ISO 9999:2002—Technical aids for persons with disabilities—Classification and terminology*. Geneva: International Organization for Standards.

Keates, S. (2005). Introducing ease of access into IBM. *ACM SIGACCESS Newsletter*, No. 82, June 2005. Retrieved January 14, 2006, from http://www.acm.org/sigaccess/newsletter/june05.php

Keates, S., and Clarkson, P. J. (2003). *Countering design exclusion: An introduction to inclusive design*. London: Springer-Verlag.

Keates, S., and Clarkson, P. J. (2003a). Countering design exclusion through inclusive design. In *Proceedings of the 2003 Conference on Universal Usability*, (10–11 November, 2003, Vancouver, British Columbia—pp. 69–76), New York: ACM Press.

Keates, S., Langdon, P., Clarkson, P. J., and Robinson, P. (2002). User models and user physical capability. *User Modeling and User-Adapted Interaction (UMUAI). 12*(2–3): 139–169.

Mahoney, R. M., Jackson, R. D., and Dargie, G. D. (1992). An interactive robot quantitative assessment test. In *Proceedings of RESNA 1992* (pp. 110–112). Arlington, VA, RESNAPress.

Martin, J., Meltzer, H., and Elliot, D. (1988). *The prevalence of disability among adults.* London: Her Majesty's Stationery Office.

Mayhew, D. J. (1999). *The usability engineering lifecycle.* San Francisco: Morgan Kaufman Publishers.

Nielsen, J. (1993). *Usability engineering.* San Francisco: Morgan Kaufman Publishers.

Nielsen, J., and Mack, R. L. (1994). *Usability inspection methods.* New York: John Wiley and Sons.

Norman, D. (1998). *The design of everyday things.* London: The MIT Press.

Nussbaum, B. (2004). The power of design. *Business Week* (17th May, 2004).

PBS. (2005). *Infant mortality and life expectancy.* Retrieved January 14, 2006, from http://www.pbs.org/fmc/timeline/dmortality.htm.

Peebles, L., and Norris, B. (1998). *Adultdata: the handbook of adult anthropometric and strength measurements—data for design safety.* London: Department of Trade and Industry.

Pfeiffer, D. (2002). The philosophical foundations of disability studies. *Disability Studies Quarterly, 22*(2), 3–23.

Pirkl, J. J. (1994). *Transgenerational design: products for an aging population.* New York: Van Nostrand Reinhold.

Popovic, V. (1999). Product evaluation methods and their importance in designing interactive artefacts. In Jordan and Green (Eds.), *Human Factors in Product Design* (pp. 26–35), London: Taylor and Francis.

Rehab Act. (1973). *Rehabilitation Act of 1973* (U.S. Pub. L. 93-112). Retrieved January 11, 2006, from http://www.dotcr.ost.dot.gov/documents/ycr/REHABACT.HTM

Rose, D. (2004). Don't call me handicapped. *BBC disability Web site Ouch!* Retrieved January 14, 2006, from http://news.bbc.co.uk/2/hi/uk_news/magazine/3708576.stm

RSA. (2005). *Welcome to inclusive design.* Retrieved January 14, 2006, from http://www.inclusivedesign.org.uk/

Semmence, J., Gault, S., Hussain, M., Hall, P., Stanborough, J., and Pickering, E. (1998). *Family resources survey—Great Britain 1996-7.* London: Department of Social Security.

SENDA. (2001). *Special educational needs and disability act 2001.* Retrieved January 11, 2006, from http://www.opsi.gov.uk/acts/acts2001/20010010.htm

Shneiderman, B. (2000). Universal usability. *Communications of the ACM, 43*(5):84–91.

Smith, S., Norris, B., and Peebles, L. (2000). *Older Adultdata: the handbook of measurements and capabilities of the older adult—data for design safety.* London: Department of Trade and Industry.

Telecoms Act. (1996). *Telecommunications act* (U.S. Pub. L. 104-104). Retrieved January 11, 2006, from http://www.fcc.gov/Reports/tcom1996.txt

UN. (2002). *World population ageing: 1950–2050. United Nations Department of Economic and Social Affairs, Population Division.* Retrieved January 14, 2006, http://www.un.org/esa/population/publications/worldageing19502050/

U.S. DoJ. (2004). *A guide to disability rights laws.* U.S. Department of Justice, Civil Rights Division, Disability Rights Section.

USCB. (2004). *American community survey.* Retrieved March 15, 2006, from http://factfinder.census.gov/jsp/saff/SAFFInfo.jsp?_pageId = sp1_acs&_submenuId

Vanderheiden, G., and Tobias, J. (2000). Universal Design of Consumer Products: Current Industry Practice and Perceptions. In *Proceedings of the XIVth Triennial Congress of the International Ergonomics Association and 44th Annual Meeting of the Human Factors and Ergonomics Society*, (vol. 6, pp. 19–22). Santa Monica: Human Factors and Ergonomics Society.

Vischeck. (2005). *Vischeck*. Retrieved January 14, 2006, from http://www.vischeck.com/

W3C. (2005). *Web content accessibility guidelines 2.0*. Retrieved April 6, 2006, from http://www.w3.org/TR/WCAG20/

Watchfire. (2005). *WebXACT*. Retrieved January 14, 2006, from http://webxact.watchfire.com/

WHO. (2001). *International classification of impairment, disability and health (ICF)*. Geneva: World Health Organization.

WIA. (1998). *Workforce investment act* (U.S. Pub. L. 105-220). Retrieved January 11, 2006, from http://www.doleta.gov/usworkforce/wia/wialaw.txt

Wong, M. (2005). Apple juggernaut shows no sign of waning. *The Mercury News/AP Wire* (Dec. 15, 2005). Retrieved January 14, 2006 http://www.mercurynews.com/mld/mercury news/news/breaking_news/13415110.htm

Zultner, R. (1993). TQM for technical teams. *Communications of the ACM. 36*(10), 79–91.

Index

task analysis, 90, 95, 97–98, 116–117
Telecommunications Act (1996), 8–9
television
    closed captioning, see subtitles
    digital terrestrial (DTT), 141–158
    digital text, 143
    electronic program guide
    integrated digital (iDTV), 142
    set-top box, 142–143, 144, 145–158
    subtitles, 143
    teletext/fastest, 143
think aloud, 138–139
ticket machine, 13
tools, 69–72
    compliance testing, 69–70
    modeling, 71
    visualization, 70–71
Total Quality Management (TQM), 88
Trace Center (University of Wisconsin-Madison),
    12, 74
training, 73–74

United Nations, 18
universal access, 43, 73
universal design, 12–13, 43, 73
usability, 70–71, 94
    definition, 3
    designing for, 4
    engineering, 69
usable access, 81–85
use-case scenarios, 146–148

user-centered design, 26, 87–89
user models, 98–99
    Model Human Processor, 124–126, 144
user mental model, 90–92, 110, 145,
    154–155
user trials, 5–7, 59, 81, 92, 94, 100, 108–109,
    129–140, 148–154
    conducting, 137–139, 151–152
    ethical approval, 130–131
    finding and recruiting for, 120–122
    location, 133–137
    preparation for, 129–133
    sampling for, 116–121, 148–149
    statistical analysis, 122–123, 152–154
utility, 61, 90
    definition, 3

VCR, 4
Vischeck, 70

washing machine, 95–96, 97–98
Watchfire, 69–70
Web Accessibility Initiative (WAI), 30–31, 41,
    70, 75–76, 78
Web Content Authoring Guidelines (WCAG),
    see Web Accessibility Initiative
Web site, 30, 53, 62, 74, 75–76
WebXACT, see Watchfire
Workforce Investment Act (1998), see
    Reauthorized Rehabilitation Act
written records, see documentation